Contact!

The book *Contact!* is an initiative of the Netherlands Institute for Social Research (SCP), My Child Online Foundation and the Dutch programme Digivaardig & Digibewust (Digi skilled & Digi aware). This book has been compiled thanks to contributions by the Digi skilled & Digi aware programme, KPN Royal Dutch Telecom (the leading telecommunications and ICT services provider in the Netherlands) and the Dutch Democracy and Media Foundation.

EDITORS
Jos de Haan and Remco Pijpers

COPY EDITOR
Henk Boeke

STEERING COMMITTEE
Wim Bekkers (NICAM), Marjolijn Bonthuis (ECP-EPN) and Regina van den Eijnden (UU)

EDITED BY JOS DE HAAN
AND REMCO PIJPERS

Contact!

Children and new media

Bohn Stafleu van Loghum
Houten 2010

Design: Nanja Toebak, 's-Hertogenbosch
Coverdesign Suzanne Hertogs (Ontwerphaven)
Photography: Hans de Bruijn, Olav Kaspers and Jasper Zwartjes
English Translation: Kirsten van Hasselt

Bohn Stafleu van Loghum
Het Spoor 2
Postbus 246
3990 GA Houten
www.bsl.nl

Table of contents

Preface   7

1 · Children online   11
Jos de Haan, Remco Pijpers

2 · Games   29
Jeroen Jansz, Peter Nikken

3 · Casual games   53
Menno Deen, Nathalie Korsman

4 · Everyday creativity in virtual worlds   69
Mijke Slot

5 · Communicating online   89
Patti Valkenburg, Jochen Peter

6 · Hyves   103
Marion Duimel

7 · Advertising   121
Esther Rozendaal

8 · Mobile phones   139
Marion Duimel

9 · Information skills   155
   Els Kuiper

10 · Education   171
   Alfons ten Brummelhuis

11 · Parental mediation   189
   Peter Nikken, Justine Pardoen

12 · Trends, conclusions and recommendations   209
   Jos de Haan, Remco Pijpers

   About the authors   223
   Contributors   229

Preface

While tidying up one of my cupboards recently, I stumbled across a box filled with articles that I had collected between 1981 and 1984 on computers. I had very neatly attached an index card to each article with a short summary. This was still possible then. After 1984 it was impossible to keep up with the changes. We were in America that year and at a university dinner one of the guests proudly arrived with a portable Apple Macintosh SE in a bright yellow bag. His demonstration generated many oohs and aahs from colleagues. The start of an unstoppable revolution.

My initial fascination had also been aroused in America in 1981, when my husband and I visited a nursery school there. Children used computers to teach them letters, numbers and shapes. Shortly after returning home my publisher, also the publisher of this book, lent me a computer. They had two and were experimenting with them to see whether authors could make the corrections more efficiently with new printing. An enormous black cube of about sixty centimetres with a 15 x 15 cm-screen. This is almost thirty years ago. I now walk around with an iPhone. Many years during which I have had to learn to live with computers, or rather, with the ever-increasing new applications of computers.

Nowadays, children take computers, mobile phones and internet for granted. This does not, however, mean that they do not have to learn how to cope with all the applications. Children used to be born surrounded by books but they still had to learn how to read and write before they could enjoy them.

This book shows what children aged between 6 and 12 have to do to become digitally literate. This is done in a level-headed way. The authors have avoided the pitfall of researchers who used to concentrate on the theme 'children and television'. The undertone then was: should-we-really-want-this-for-our-children? But there was – and still is – nothing to want. Television came, never to go away again. The same applies to new media that, as mentioned in the introduction, are not new for children at all but the most normal thing in the world. "New" for one our grandsons was the gramophone that he found in the attic. When played it caused as much amazement as the small Apple did at the time among the American professors.

The book gives a clear state of affairs, based on research, of what is known about what children can and cannot do, do and do not do, do and do not want to do. Included are striking examples of what children themselves say about this. Misunderstandings are dispelled. For example, the misunderstanding that children generally have more ICT skills than their parents. Young children now have parents who have been confident for a while using computers. These parents also use Hyves and are, besides classmates, the most frequent friends on children's Hyves. And 30% also have a grandmother or grandfather in their social network. What do you mean 'new media'?
At the same time, the idea that children automatically make optimal use of digital media because they are so skilled in making use of it, is refuted. No, they have to learn this just as they have to learn other life skills. And for this they need adults, who can use this book to find the necessary information to do this.

All aspects of using 'new media' are discussed, from gaming to instant messaging and from mobile phones to participating in virtual worlds. Specific knowledge is used to describe the many possibilities and the few risks of modern communication media. In almost all the chapters it becomes clear that children mainly care about keeping in touch.

In my collection's oldest cutting, a mathematics teacher tells how she took her pupils to Wageningen University to familiarize the children with computers. Answering the interviewers question "Is the computer not the start of a lonely era?", she said: "On the contrary, once the computer has become fully integrated into everyday life, communication with each other will become much easier." And that is exactly what has happened. Although there are still people who do not want to believe this.

Rita Kohnstamm

1

Children online

Jos de Haan
Remco Pijpers

Ask a child to imagine a life without new media and images of boredom loom up. "I cannot understand how my mother used to enjoy herself. What did she do without a PC, without internet, without MSN? She must have been bored stiff?" These are Tara's words, aged 15, who was cited in the ICT and Society Yearbook 2006: The Digital Generation.[1] This book – as so many others lately – deals with teenagers. Children aged between 6 and 12 are paid very little attention. But these children too take a life with game consoles, mobile phones, PCs and internet for granted. Reason enough to delve deeper into the media use of this group.

1.1 • WHAT ARE WE TALKING ABOUT?

Nowadays, new media are a source of entertainment for the youth, as well as an essential link in communications. They play *casual games* (simple games) on the internet, they play *multiplayer games* on game computers connected to the internet, and they play in virtual worlds such as Stardoll.nl and Habbo. On the video site YouTube they watch films, sometimes made by themselves and even bringing international fame to some. They present themselves on Hyves and 'scribble' compliments on each others pages, on photos taken on social networking sites such as Facebook and Hyves by their digital cameras or mobile phone.

After all, they use their phone for much more than just making phone calls. They declare their love for each other and have arguments on MSN and via text messages.

In other words, for children and youngsters 'new media' are part of everyday life. For them, they are no longer really new media. In the same way that for us adults airplanes are no longer a 'new means of transport'.

Parents and teachers sometimes react amazed – or with admiration – at the ease and speed with which young people master the digital life at such an early age. But they are also concerned. Parents are sometimes worried about the amount of time children spend on the computer, they are concerned about bullying on MSN or Hyves, or they are shocked by violent or pornographic images seen by their children. And teachers regularly complain about the uncritical cutting and pasting of information from the internet for school projects.

What do parents and teachers do about this? Usually not a lot. Partly because there is so little information available. That is what this book hopes to change.

It is not just parents and teachers who are struggling with the media use of children and young children. Innumerable trend watchers, journalists and columnists have meanwhile expressed their ideas on the advantages and dangers of new media. A distinction can be made between the optimists and the pessimists.

The optimists – in the Netherlands also referred to as the 'Einstein Generation' followers – attribute young media users unique qualities, precisely the way futurologists sang praise in the 90s of the new internet generation. Modern children would be capable of everything digitally, easily 'outsurfing' adults.[2] An astonishing acceleration of the evolution means that they can now suddenly multitask (listen to music and watch TV and use MSN and do their homework at the same time) and they keep getting smarter.[3] With the mobile internet, the double-quick 'Wifi-generation' would be able to make use of the digital possibilities anytime, anywhere.[4]

The pessimists are particularly prone to pointing our their objections. They feel that children lack information skills and cannot recognise advertisements. Children are forever bullying, stealing each other's virtual identities and are said to be susceptible to cyber lurers.

The truth, no doubt, lies somewhere in the middle. But this does not solve the problem. What do children have to learn to be able to handle new media in a safe and efficient way? Which skills are needed to transform them into media-literate citizens? And what is the task of their carers?

Because of the focus on teenagers, little scientific research is available on the use of new media by 6- to 12-year-olds. But is not totally unavailable. We wanted to take stock of the limited research available and have therefore put together this book. The following main questions were asked :
• What is available digitally for children aged between 6 and 12?
• How do they use digital media?
• Which skills have they mastered and which do they lack?
• What are the opportunities and what are the risks for these children?

1.2 • HOW THE INTERNET BECAME SUCCESSFUL AMONG CHILDREN

At present, most children assume that the internet 'is just there', comparable to electricity from a socket. They can be sincerely surprised at times when the internet is not or not yet available, say for example, in a car. This is totally new for them.

The wide distribution of internet is very new. It was not until the second half of the nineties that the diffusion of internet in households in most western countries really took off. In 1997, only one in every three families in the Netherlands had a 'fast' computer, with a 486 or 586 (Pentium-) processor. Another third had to cope without a computer. And if there was a computer in the house at all, it was used as a word processor or to 'learn school subjects', as Qrius' research among youngsters showed.[5] In those days, children aged 6 to 12 were mainly busy playing games, either on a PC (with a CD-ROM) or on the Gameboy, the Nintendo *handheld* of that time. The IJsfontein company, nowadays specialised in websites for children, started out in 1997 as a producer of CD-ROMs for children. That market was promising. And television was still, by far, the most popular medium.

About 10 or 12 years ago, analog modems were used and it was necessary to 'dial-up' to the internet. After many beeps a connection was made allowing surfing and emailing. The information was mainly text based, images were the size of stamps and there was no or hardly any sound, let alone video. In those days this was unappealing for children – who were used to TVs and CD-ROMS. Teenagers at that time had 'problems with the unorganised character of the internet and did not exactly understand its use'.[6] The internet was then not yet a mass medium: merely 7 to 8% of Dutch children and young adults (aged between 6 and 19) had a PC with internet at their disposal, and as many as 48% of the 6- to 8-year-olds had never heard of the internet. Despite this, young people were redubbed as the 'internet generation' in 1998 by Don Tapscott, in other words, the generation growing up with internet computers who would teach themselves to become skilled users.[7]

Not until the breakthrough of broadband internet (ADSL and cable, after 2000) did the internet become really appealing to children and young adults. The Netherlands quickly became a broadband country. It was ahead of other countries as it already had a good cable television and telephone line infrastructure present that could be made suitable for fast internet.

The competition between cable companies (with internet via cable line) on the one hand and ADSL providers (with internet via telephone lines) on the other, caused a drop in prices and resulted in a rapid increase in the number of broadband households. By 2000, one in every ten people in the Netherlands had access to the internet via a cable connection and in 2001 the advance of ADSL started. Initiatives such as 'fibre-optic cable to the front door' were carefully continued.

In 2009, 77% of the Dutch households already had broadband connections, 13% had another internet connection and 10% was not yet connected.[8] Families led the way in broadband distribution via ADSL and 'the cable'. Flat fees meant that children now had unlimited internet access without their parents having to worry about high telephone bills. In the year 2010, almost all children and young adults are permanently online.

But not only broadband and flat fees made the internet appealing to children and young adults. Other developments also contributed to its success, namely:
· **The rise of communication services such as MSN** (now called Windows Live);
· **The possibility of adding your own information to the web** (web

2.0); children can write their own articles online, upload photos or put films online. Creating this so-called *user generated content* is also facilitated on social networks, such as Hyves, which was founded in 2004.

- **Convergence of internet, telephone and audiovisual media**; this is making the differences between television, computer and telephone disappear. Children can watch television programmes on the computer, surf on the internet with the latest phone and vote for candidates in TV shows via text messages.
- **Increasing availability of content**; the supply of information (text, sound and image) has grown sharply resulting in everybody being able to access information and entertainment through various media, anytime and thanks to the mobile media, anywhere.

This particularly involves new applications. Computers, as such, are no longer new, nor are the internet and mobile phones. We specifically refer to new applications when we talk about 'new media'. Really new is the mobile internet, about to make inroads. The iPhone is a forerunner of this.

1.3 • NEW TREND: ONLINE AT INCREASINGLY YOUNGER AGE

EU Kids Online, a project that charts all European research on young people and the internet, shows a new trend: children are starting to use the internet at an increasingly younger age. Furthermore, they make more intensive use of the internet.

In 2008, 86% of the European 15- to 17-year-olds were online. At that time all Dutch, Swedish and Finnish children in this age group had access to the internet. This is 100%. At home, at school or at their friends. For the 11- to 14-year-olds it was 84% of the European children and 96% of the Dutch children and for the 6- to 10-year-olds it was 60% in Europe and 83% in the Netherlands. The Netherlands, together with the Scandinavian countries, stands out as one of the leaders with regard to internet penetration. This book for the most part focuses on the Netherlands because of its leading position and the opportunity to study the chances and risks of internet use in a situation of near market saturation. It will not take long before all Dutch 6- to 12-year-olds are online. Furthermore, the number of 6-to 12-year-olds who own a mobile phone is also growing rapidly. This trend of using new media at a younger age is visible in all European countries.[9]

Online activities that mainly appealed to youngsters until now, are increasingly being taken over by children. Hyves, for example, was initially set up for students but a growing number of primary school pupils now also join in. And although Habbo Hotel was meant for teenagers aged between 12 and 18, in reality, many much younger children take part.

At the young age of 6 children are already starting to play casual games (simple games on the internet). The number of games on offer is rapidly increasing. There is, for example, *The Littlest Pet Shop*, a virtual world full of games for children aged from 4 onwards, which can be accessed using a code. The code is on the inside of the collar of a cuddly toy that can be bought in shops for €15. And in the *Club Penguin* children aged older than 5 can furnish an igloo and invite others so that they can talk together using pre-programmed sentences and emoticons.

1.4 • DEVELOPMENTAL PSYCHOLOGICAL BACKGROUNDS

A previously mentioned, this book focusses on the use of new media by 6- to 12-year-olds. The age does not just mark a school period but it also relates to the developmental psychology.

Fantasy and reality often overlap for pre-schoolers. What the pre-schoolers (aged 4 to 5) see on television, hear in fairy tales or experience in a computer game is as real for them as what they see around them. A frog can become a prince and Santa's sleigh can fly through the air. No problem.

From the age of 6 the ability to think has developed to such an extent that a child can distinguish between reality and fantasy, although they can co-exist. The child can see that Santa Claus wears the neighbour's shoes but does not see this as a problem.

According to developmental psychologists – for example, the Swiss psychologist Jean Piaget – children aged between 6 and 12 find themselves in the concrete operational stage of cognitive development. At this stage the child, for example, learns to compare lengths and volumes, to sort things, count, do sums and think figuratively.

In addition, the identity development also plays a part in this period, whereby children see themselves in relation to others. Important work in this field was carried out by the American psychologist Erik Erikson. He is convinced that children do not just learn cognitive skills between the ages of 6 and 12, but also develop self-confidence in this period, assuming that positive behaviour is sufficiently rewarded and encouraged (negative feedback can give children an inferiority complex).[10]

During the teenage years, children enter a new development stage in both their cognitive and their social growth. According to Piaget the formal operational stage starts at the age of 11 and this is the phase in which they learn inductive and deductive reasoning and to grasp abstract concepts. At the same time, puberty starts around this age, accompanied by all its sexual interests and attempts to 'belong'.

This knowledge on the psychological development helps understand what media 'does' to children and to helps to determine what carers should do (or not do).

1.5 • FOUR THEMES

As mentioned, this book is based on the following main themes: the availability of websites, digital content and communication tools for children aged between 6 to 12, their use of these digital media, the skills they master and the skills they lack, and finally the opportunities and the risks for these children. A clarification of the respective themes can be found below:

DIGITAL AVAILABILITY

In 1998, KidsPlanet was one of the first professional Dutch children's websites. Internet provider Planet Internet (later changed to KPN) wanted to provide children aged between 7 and 12 with a safe, friendly spot on the internet with this website. The jury of the Gouden @penstaart (award for the best children's website) voted KidsPlanet as the best children's site of the year in 1999. A quote from the jury's report: "You can actively participate and it is very colourful". Active participation at the time meant that you could play a game, take part in a competition, print a colouring picture and send an email to the editors. In fact, nothing more than an old-fashioned magazine, but digital. It was 'old-style interactive'.

Figure 1.1 • Children can think up their own game with Game Studio and build it. When others visit the site they can play their game.

A lot has changed since then:
- the number of websites for children has increased strongly;
- static content sites (with much reading material and little interaction) have made room for flashy multimedia sites with omnipresent interaction;
- statements can no longer be separated, but have become cross-media (TV programmes, plus magazines, plus games, plus websites, plus weblogs).

Although there are enough attractive children's sites, it is striking that most do not meet the criteria for a good children's sites, such as appeal, reliability and user-friendliness.

THE USE OF NEW MEDIA

Old media, such as radio and television, are one-way traffic. You can absorb information but cannot add anything yourself. The internet has radically changed this. Children are increasingly becoming their own producers of information. They can, for example, make their own cartoons (among other, on the Nickelodeon site) rather than just watching

Figure 1.2 • You can make your own cartoons on the Nickelodeon.nl site.

what professionals have put online. This is what is called content creation (creating your own text and images).

One-way communication has been replaced with 'real' communication:
not just with the makers of a site but also among the users themselves.
This is particularly appealing to children. This is why social networks,
profile sites and multi-user games have become so popular.

A third aspect of the use of new media is that communication and playing show more overlap. The most popular games are those in which you
can communicate with each other. For example, Runescape for boys,
Neopets and goSupermodel for girls, and Habbo Hotel for boys and
girls.

SKILLS

Which digital skills have children mastered and which do they still lack?
Can they teach themselves – with the greatest of ease – the use of new
technologies or do they require assistance? What role do parents and
teachers play here?

In 2005, the Dutch Council for Culture introduced the term 'media-
wisdom' (in other countries mostly referred to as media literacy) which
has received much attention since then. The Council's definition of me-
dia-wisdom is 'the combination of knowledge, skills and mentality enabling citizens to consciously, critically and actively take part in a complex, changing and fundamentally 'medialised' world'.[II]

Being media-wise (media literacy) entails knowledge, skills and mentality:
• **knowledge** – the historical and ethical knowledge necessary to be
  able to interpret and assess media messages;
• **skills** – the technical and creative skills necessary to be able to use
  media;
• **mentality** – the willingness to use the knowledge and skills to increase their own awareness (and the awareness of others).

How should these concepts be translated into what children do on the
computer? This has been extensively discussed but little empirical research has been carried out. The articles in this book present the current
state of affairs.

OPPORTUNITIES AND RISKS

In order to chart the opportunities and risks, we have used the classification developed by the EU kids online project. The categories are:
• managing information (content);
• communication with others (contact);
• own initiatives (conduct).

These three ways of dealing with new media can take shape in various
areas, including education, creativity and identity development.

Proper use of IT in education has a positive effect on the learning
achievements[12] and may increase the motivation. Teenagers say that if
they understand ICT lessons better, they learn more quickly and find
education more interesting and more fun.[13] But does this also apply to
6- to 12-year-olds?

Active and independent learning are becoming increasingly important.
This kind of learning often replaces traditional knowledge transfer in
class.[14] Whether this is good or bad is not something we will be dealing
with, but it is a fact that pupils have to look for information, for say a
school project, more and more often.[15] This increasingly involves the
use of educational games, video footage, online audio, etc. [16]

In short: enough opportunities. But there are, of course, also dangers.
Both aspects are discussed in this book.

1.6 • READING POINTERS
• **Chapter 2** (Jeroen Jansz and Peter Nikken) is about games for 6- to
12-year-olds. What kind of games do these children play? The authors
observed that boys and girls are equally enthusiastic players but that
genre preferences are obvious at an early age. They also deal with what
it is that attracts children in entertainment games. They also establish the
positive and negative effects of the games that children of this age play.

• **Chapter 3** (Menno Deen and Nathalie Korsman) is about casual
games. The authors describe which games are played, for example,
shooting games, dress-up games and puzzles. The quality of these
games appears to differ greatly. Apart from nice games you also stumble
across terrifying and sexually tinted games, while the visitors are not
even asked their age.

• **Chapter 4** (Mijke Slot) deals with virtual worlds for children among which Habbo Hotel. But also: GoSupermodel, Runescape and Neopets. She pays special attention to the roles that children play in the virtual worlds, and the way in which the 'community' stimulates creative interaction. How do children express their creativity ?

• **Chapter 5** (Patti Valkenburg and Jochen Peter) shows how important online communication is for children. Children are especially attracted to instant messaging, in particular, MSN. The authors pay special attention to the opportunities offered by online communication to achieve intimacy and contribute to identity development, social skills and self-confidence. They also discuss the possible risks of the internet, such as online bullying, sexual intimidation and privacy risks. Finally, the authors make recommendations for stimulating positive use and reducing the dangers of undesirable consequences.

• **Chapter 6** (Marion Duimel) is about Hyves. Until recently, MSN was most popular among children and young adults but now it is Hyves. It is *the* network site in the Netherlands, initially intended for young adults but along the way an increasing number of young people joined and now many primary school pupils also have a profile. The author presents a broad exploration of the Hyves profile contents of children up to the age of 12.

• **Chapter 7** (Esther van Rozendaal) is about advertising on the internet. Children see banners and buttons, but they are also brought in as brand ambassadors. What kinds of advertising do children see on the internet? How do children process these new advertising forms? The author also plunged into the world of online trade and has some fascinating examples.

• **Chapter 8** (Marion Duimel) describes how active 6- to 12- year-olds are with mobile phones. Many parents find it comforting to know that their child can be reached by mobile phone. Phones are, however, becoming increasingly advanced and can now, in addition to calling and text messaging, also play music, take photos and films and surf the internet. To what extent do children connect to these new possibilities? Is the phone just an extra communication means with their parents or does it also serve as a social channel with friends? What rules do parents set?

• **Chapter 9** (Els Kuiper) is about digital natives, in other words, children born into a world riddled with digital media.[17] The impression is often created that this generation can manage ICT applications by 'themselves'. Scientific research, however, shows a different picture. This chapter raises the matter of the internet skills (and the lack thereof) of children aged between 8 and 12.

• **Chapter 10** (Alfons ten Brummelhuis) deals with the question "What does education do to increase the information skills of pupils?" What do we know, for example, about the effects of the use of new media on the development of the learning achievements of children? What are the changes still necessary for education to sufficiently prepare children for a role in a knowledge economy?

• **Chapter 11** (Peter Nikken and Justine Pardoen) gives an overview of the scientific research relevant for parents. The truth about the cliché image that parents lag behind their children. And what should they really know to be able to guide their children?

• **Chapter 12** (Jos de Haan and Remco Pijpers) closes with conclusions and recommendations, considering the opportunities and risks closes with a description of trends and with conclusions.

A book about children and new media would not be complete without introducing those involved. Journalist Lotte Boot and photographers Olav Kaspers and Hans de Bruijn made portraits of nine children, a mother, and a primary school teacher. Suzanne Hertogs of Ontwerp-haven made the illustrations.

NOTES

1   Haan, J. de & Van 't Hof, C. (2006). *Jaarboek ICT en Samenleving 2006*: de digitale generatie. (ICT and Society Yearbook 2006: the digital generation) Amsterdam: Boom.

2   Boschma, J. & Groen, I. (2006). *Generatie Einstein: slimmer, sneller, socialer.* Communiceren met jongeren van de 21e eeuw. Einstein Generation: smarter, faster, more social. Communicating with adolescents of the 21st century) Amsterdam: Pearson Education.

3   Johnson, S. (2005). Everything *Bad is Good for You; How Today's Popular Culture Is Actually Making Us Smarter.* London: Allen Lane.

4   Delver, B. & Hop, L. (2009). *De WIFI-generatie;* de jeugd op het mobiele internet. Vliegensvlug en Vogelvrij. (The WIFI generation; adolescents on the internet, Quick and Free) The Hague: Van Stockum.

5   Qrius (1997). *Jongerenonderzoek 1997.* Research on Adolescents Amsterdam: Qrius,

6   Qrius (1997). See note 5.

7   Tapscott, D. (1998). *Growing up digital*; the rise of the net generation. New York: McGraw-Hill.

8   CBS (2009). *De digitale economie 2009.* (The digital economy 2009) The Hague/Heerlen: CBS. See p.137.

9   Livingstone, S. & Haddon, L. (2009). *EU Kids Online: Final report.* (EC Safer Internet plus Programme Deliverable D6.5). London: EU Kids Online. p.5.

10 Piaget, J.-P. (1952). *The origins of intelligence in children.* International Universities Press, New York.
    Erikson, E. (1950). *Childhood and Society.* New York: W.W. Norton. (Eriksons basic work).

11 Raad voor Cultuur (2005). *Mediawijsheid: de ontwikkeling van nieuw burgerschap.* (Media literacy: the development of new citizenship) The Hague: Raad voor Cultuur. Definition on p.8.

12 Balanskat, A., Blamire, R. & Kefala, S. (2006). *The ICT impact report.* A review of studies of ICT impact on schools in Europe. Brussels: European schoolnet.
    OCW (2006). *Verbonden met ICT* (actieplan). (Connected with ICT) The Hague: OCW.

13 Dialogic (2006). *Breedband monitor.* (Broadband monitor) Utrecht: Dialogic;
    Brummelhuis, A ten (2006). 'ICT heeft leraar hard nodig'. (ICT needs teachers) In: *Vives*, January 2006;
    Vier in Balans Monitor 2007 (Four in Balance Monitor), *Stand van zaken over ICT in het onderwijs.* (ICT state of affairs in education) Kennisnet ICT at school. See www. ICTopschool.net.

14 Gennip, H. van & Braam, H. (2005). 'Leren en ICT. In: J. de Haan & L. van der Laan (eds.), *Jaarboek ICT en samenleving 2005 (ICT and Society Yearbook 2005).* Kennis in netwerken. Amsterdam: Boom. p.129-146.

15 Blauw Research (2005). *Internet in de klas.(Internet in the classroom)* Rotterdam: Blauw Research bv.

16 Dialogic (2006). See note 13.

17 Prensky, M. (2001). Digital natives, digital immigrants. *On the Horizon* Vol. 9 No. 5, October 2001.http://www.marcprensky.com/writing/Prensky-DigitalNatives, DigitalImmigrants-Part1.pdf.

Enjoying the Wii together

The houses of the brothers Milo (6) and Iloy (11) from Baarn, and their neighbours Beau (8) and Elvis (10) are invaded daily by various neighbourhood children. Together, they have a whole range of game consoles: a Wii, a PSP, a PS II, a GameCube, four Nintendo DSs, and two Xboxes.

Iloy: "We almost always play together. We Wii in our living room, for example, and Beau brings along his own remote control. We often play competitions in teams, two against two."

Beau: "I've got a DS and an Xbox 360. I play racing games on these, and shooting games and fighting games. I used to play for hours, but I'm no longer allowed to. I now have to play outside. That can be fun too."

Milo: "I often watch when they're playing on the Wii, I don't want to play too often myself. I then play on their DS, for example, Barnyard, that's easier. I like it when older children come to play here. Outside I'm not allowed to play with them but inside I can join in and learn from them."

Elvis: "I really enjoy it. We have crisps and soft drinks. And our mothers have wine and nuts. You also have a greater chance of winning if you all play together against the Wii. We sometimes argue about whose turn it is. We then have to switch the Wii off and go and play outside. Hide and seek or shooting each other with toy guns."

Beau: "We've got those guns with foam bullets and we've now got great hiding places and know how to shoot because we've played those shooting games."

Iloy: "Yes! We first send someone to divert him and then someone else shoots him from the back."

Elvis: "I'm best at that. I climb into a tree and snipe from there. I've got a DS and a PSP. I often have games rated 18. I don't find these difficult but some are scary. I hear something behind me and when I turn around it's a monster. In these fights everything starts to tremble. Sometimes I then turn away."

Milo: "I also sometimes find it scary when Iloy plays. Then I shout: Watch out! Behind you!' I sometimes crawl under the blankets until it's over."

Iloy: "I'm never fed up. I'd like to become a game tester or a game maker."

Milo: "I want to work in a zoo! I love animals. Especially the gazelle."

Marianne Drisdale, Iloy and Milo's mother: "They are not allowed to play games rated 16+ when Milo is also there but apparently they sometimes do this secretly. Milo hadn't told me that he's scared of certain games. I'll have to watch out for this even more than I already do. One of my rules is that they're allowed to play games for a maximum of an hour and then they have to go outside. But sometimes they then go and play on Beau's Xbox. Particularly Iloy is very fanatical, I really have to slow him down. We mothers therefore check whether the boys have been playing games at another boys' house. It's great fun though when the whole neighbourhood is here. They really enjoy themselves together. They are reasonably free and easy when they are here, as long as they are nice to each other. So they like coming here. They're a noisy lot when they're here and it's not just these four then. But I don't mind a lot of noise."

2

Games

Jeroen Jansz
Peter Nikken

Koen (11) quickly finishes his soup and gets up from the table. "Where are you going?" asks his father. Koen says that he has to rush upstairs to go online because his Warcraft clan *has agreed on a raid in 3 minutes. "No way. You'll stay here and finish dinner with us". Koen stares at his father surprised and stammers: "I have a social life too you know."*

Video and computer games comprise a unique segment of entertainment media as it offers young children an exceptional combination of possibilities. Players can choose from numerous titles and genres, whenever they want. They themselves choose the fantasy world they want to enter and what they will and will not do there. First they enjoy themselves trying to unravel a complicated mystery, then they have fun at an even higher game level by solving the problem. In addition, online games, such as Koen's *World of Warcraft*, make it possible to challenge another team with your group of friends, anywhere in the world. Victories are proudly celebrated. But even losing can have its advantages: sharing disappointments strengthens the bond between friends.

The appeal of video and computer games is mirrored in their commercial success. The Dutch games market grew spectacularly over the past five years, from € 187 million (sales) in 2004 to € 342 million in 2008.[1]

Figure 2.1 • Screenshot from *World of Warcraft*. Online games make it possible to challenge another team with a group of friends, anywhere in the world.

This growth is comparable to the increase in sales in the United States and the United Kingdom.[2]

HARDWARE

In the 70s, games were almost entirely limited to playing on large game machines in so-called arcades. In the course of the 1980s the emphasis shifted to homes. Here one played on personal computers or a special game computer, a so-called console that was connected to the television.

Over the last years, the console games market has grown faster than the market for PC games.[3] This is mainly caused by the rapid developments in hardware. The *Playstation* consoles (PS2 and PS3) made by market leader Sony, are experiencing fierce competition from Microsoft's XBox360, while Nintendo appears to have achieved its own market niche with its Wii. The PC, however, continues to be an appealing game platform for many because of its obvious access to games on the internet, although playing online games from the PS3 and Xbox is now also growing rapidly.

The third game platform is comprised of mobile hardware. In 1989, Nintendo launched the GameBoy which quickly reached a mass audience of young players. Success was consolidated with the Nintendo DS, followed by the DSi, despite competition from Sony's Playstation Portable.

Recently, the popularity of games on mobile phones has grown. The games and java games were initially very easy but internet access on many phones has enabled an increase in the choice of games.

THE IMAGE

You would think that the wide spread of games would mean that gaming has become an accepted way to spend one's time, just as television is. But this is not the case.

The image of games in the media is still dominated by exceptional, often violent titles (such as *Manhunt* or the series *Grand Theft Auto*). In the aftermath of gruesome events, such as the murders in the German Winnenden, where a 17-year-old killed many innocent people in 2009, the media is quick to point to games as a responsible factor. The call for measures then lumps all games together.

What is usually forgotten is that most games are not violent but can be played by 'all ages', without harmful effects. Only 17% of the market is made up of games with a violent content (PEGI classification '16' or '18').[4] The types of games available are at least as varied as for films, with certain titles specifically intended for an older audience.

CONTRIBUTION TO THE DISCUSSION

This chapter intends to make a balanced, scientifically based contribution to the discussion on gaming by children aged around 12:
• First of all we outline the playing behaviour of boys and girls;
• We then answer the question *why* they play, based on theories on the appeal of games for this age group;
• We then go on to discuss the main negative and positive effects determined by research;
• Finally, we place the playing of games among 6 to 12 year olds in a broader perspective, particularly focusing on the role of the parents.

Recent research has shown that the average child aged between 6 and 12 years regularly plays a game, and increasingly often online. Dutch children do not differ much from children in other countries in this respect.

One of the first extensive Dutch studies on gaming by children showed that in 2002 already 59% of the children from the age of 8 played a game every day and 31% a few times per week.[5] At the time, mainly stand alone games on PCs and game consoles were played. The survey published by Stichting Mijn Kind Online (My Child Online Foundation) early in 2009, showed that all children (100%) between the ages of 6 and 12 played so-called 'casual' games on the internet (for example, *Bejeweled*).[6] More than 40% of the children play the more complicated *online multiplayer games*, such as Runescape and World of Warcraft.

However, besides the internet, other platforms are also still popular:
• game console (62%);
• handheld (70%);
• PC (56%);
• mobile phones (14%).

The results of British research are similar: in England, too, 100% of 6- to 8-year-olds play games on different platforms.[7] In England the mobile phone – 64% – is an even more popular platform than in the Netherlands. Moreover, British children say that they prefer games to other entertainment media, such as television.

Information on the playing behaviour of children younger than 6 years of age is limited. For the time being, it looks as if they play much less than older children as they do not yet possess all the skills required to play many of the current games. They cannot, for example, read yet. Nevertheless, it appears that 4% of the American children aged between 0 and 3 play a game on an average day and this increases to 16% for 3- to 6-year-olds.[8] It is possible that the gaming behaviour of very young children may also increase in the future if and when a larger selection of games becomes available to them.

The amount of time spent on gaming among American children increases with age. Children aged between 4 and 6 play for about 10 minutes per day, for 8- to 10-year-olds this increases to more than an hour.[9] The researchers would like to point out that the time for the youngest players is estimated by their parents. It is possible that they have underestimated the time spent playing games as they are actually not happy with the playing behaviour of their children or because the playing takes place out of their field of vision, in the child's room, for example.

The parents who took part in the first extensive Dutch study (see above) estimate that their children aged between 6 and 9 play slightly longer than one hour per day. For the 10- to 12-year-olds this was almost two hours.[10] The figures coincide with more recent research.[11]

One usually thinks of boys playing games but all research has shown that girls up to the age of 12 are also enthusiastic players. Boys of this age do, however, play more games and more often than girls.[12] There are signs that it is especially the frequency that differs: the average boy plays every day while the average girl plays a few times per week.[13]

2.3 • THE APPEAL OF GAMES

Playing computer games is very appealing to young children because it is, in fact, just 'playing'. Playing is an essential part of childhood and of great importance for a healthy development.[14] The entertainment media offers children the possibility of actually or virtually trying out, through games, how you deal with the difficult sides of existence.[15] This particularly applies to entertainment games in which different roles and perspectives can be played. Besides, repetition, characteristic of every game, is an important element. Games are especially suited to continually try out new things.

How children use games for their development depends to a large extent on their age. Already in the 1930s, Piaget indicated that the playing of children can be divided into different stages that are determined by their level of cognitive development.[16] In the current study of children and media, the cognitive development theory is widely accepted. Children are considered to be active processors of media content. The way in which they process the information depends on their development.

Children's media preferences can generally be explained by the *moderate discrepancy hypothesis*.[17] This means that children prefer media content that is slightly too difficult for their own level of development. This hypothesis agrees with what the Russian psychologist and philosopher Lev Vygotsky calls the 'zone of proximal development'; children want to learn and are therefore looking for things they *can get* a command of.[18]

The theory development of the way in which children of different ages handle games, is, however, still in its infancy. This has resulted in us partly basing our information on generalisations made in studies on watching television.

UP TO 5 YEARS OLD

The youngest children, up to about 5 years of age, have a great preference for media content with striking visual and auditory characteristics. They like to watch familiar surroundings in which events unfold slowly and are repetitive like in Teletubbies, for example, otherwise the discrepancy with their cognitive abilities is too great.[19]

American research has shown that children of this age really only want to explore (within the virtual games world).[20] These children are too young to grasp the rules of games. They master a game by trial and error and are particularly fascinated by the music, bright colours, nice graphics and the appealing cartoon characters.[21]

5- TO 8-YEAR-OLDS

After the age of 5, a preference for more challenging media content and a higher pace grows. Children now have a longer concentration span for the actual game. They enjoy unexpected events on the screen, provided they arise within a familiar context.[22]

The budding preference for action and cartoon violence also probably stems from this. Piaget's observation that children between the ages of 5 and 8 understand the rules but prefer to play by themselves, explains the popularity of Nintendo's Mario series and the fast online games.

8- TO 12-YEAR-OLDS

Taking into account that children can distinguish between fantasy and reality from about the age of 8, fantasizing changes as a result of the media.[23] Children aged between 8 and 12 have a preference for fantasizing about things that could really happen.

It is known that children like to compare themselves with characters they see on television. Children project themselves in their idols from

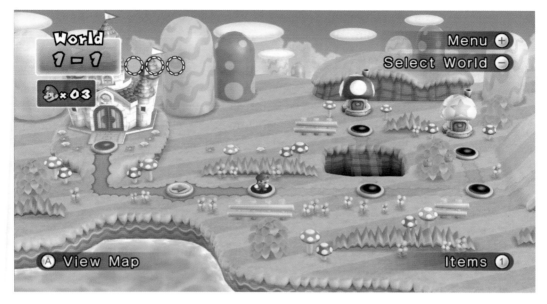

Figure 2.2 • Children aged between 5 and 8 have a strong preference for cartoons and action games such as Nintendo's Mario.

the world of music, film or sports or in less famous media characters, if they resemble them psychologically. They particularly choose characters of their own gender, although boys are stricter in this than girls.[24]

It is questionable whether children up to the age of 12 identify with game characters. On the one hand it is self-evident as they have to take on the role of the leading character as a player, but on the other hand, game characters are often flat characters with a feeble psychological profile. Not much research has been carried out yet. Only in an American study it was determined that primary school children compare themselves to game characters they like. This results in the fact that many children see themselves as less good-looking, competent or fun than the character they imitate.[25]

GENDER

The way in which children use games is not only dependent on their age, but also their gender. The game preferences of boys and girls differ from a young age and are in keeping with sex-typed preferences for toys. Boys enjoy action and combat games, adventure and sports games while girls prefer puzzles, Nintendo-like platform games and games focusing on beauty.[26]

The social role theory explains preference according to sex and the fact that boys play games more often and for longer periods, across the board, than girls.[27] The theory states that socialization of men and women into gender roles is based on their differing physical capacities. Western societies have a division of labour according to gender that has been based on this and go hand in hand with stereotypes and standards for the gender roles. Female characters are underrepresented and are usually pictured with emphasis on their female shape.[28]

The gender role explanation is confirmed by the successful sales of games that closely fit in with the stereotype female role, such as Barbie Fashion Designer of the 90s, or games about fashion and beauty that have since been marketed. Despite these successes, the dominance of male stereotypes and standards in the games world has not been broken. The game culture is, therefore, still less appealing to girls, especially those approaching puberty.[29]

Two qualitative studies have shown that the dominant culture can be broken locally. Preferences for games and playing behaviour are embedded in gender relations but are also determined by the context in which the games are played. In an after school game club for girls, for example, playing different games in the same room led to hilarious enthusiasm.[30] The girls spent an increasing amount of time playing and also chose titles normally belonging to the boys' domain. In the other study among children aged between 10 and 12, in a mixed game club, it appeared that the communication between the children when playing the online game Whyville was limited to their own gender. But if something had to be learnt or figured out, both the boys and the girls crossed the gender boundaries without reservations.[31]

The differences between the game preferences of boys and girls can also be explained using neuro-psychological factors. Most games require skills that boys are better at than girls, like mentally rotating objects, navigating through a maze and aiming a weapon. There are far fewer games that call for the better language skills of girls or their memory skills. The neuro-psychological and social role explanation are in line with each other as they both show that game preferences are embedded in structures that are wider and more solid than the behaviour of individual boys and girls.[32]

In addition to age and gender, individual differences also play an important role in the playing behaviour of children. They can choose from a broad and varied media selection to entertain themselves and some games appeal to particular groups of children more than to other children.

Within the media study, the 'uses and gratifications' approach offers an explanation for media use and media preferences.[33] The starting point of the theory is the active role of children as brought forward by the cognitive theory mentioned above. Children choose a medium, or specific media content (uses) because they expect it to satisfy specific needs (gratifications). It is about what children do with media, rather than what media do with children.

The 'uses and gratifications' theory is particularly suited to throwing light on the use of games as this medium, after all, exists thanks to the active participation of the players. Researcher Sherry and his colleagues carried out research on a group of gaming children aged about 10 and a group aged about 13.[34] In both groups the main motive to play games among both the boys and the girls was 'the challenge', followed by competition, diversion, arousal, fantasy and social interaction. The researchers also observed that a high score for 'competition', 'fantasy', and 'challenge' among 10-year-olds is related to spending a lot of time on gaming.

Furthermore, the boys also scored higher here on all the motives than the girls, but the order of ranking of motives was the same for both genders. Among the youngest players, the motive 'being stronger than in real life' was also mentioned and the social interaction motive was less important. This could indicate that 10-year-olds play especially to be intellectually challenged and to satisfy a need for fantasy.[35]

Also in a Dutch study carried out among 300 children aged between 7 and 12, it was apparent that the challenge was clearly an important element of the games.[36] The young players were asked what was important in a good game. The challenge of being able to finish a game ranked number one, followed by the possibility of being able to determine the course of the game themselves, presence of action and violence, good graphics and sound, and the possibility to learn something. Boys were considerably more convinced of the importance of the first three characteristics than girls.

Another way of gaining insight into the game preferences and needs of children is by researching the games they make themselves. At the end of the 90s, a project was launched in Boston for boys and girls aged around 10, to build their own simple games. It appeared that boys often built their games around an action theme and designed rather violent feedback on 'failure'. They used their previous game experience by including elements from commercial entertainment games. The girls lack this experience and derived their game elements from their daily lives rather than from the gaming culture.[37]

The most recent 'Girls Creating Games' project continues in the tradition, but is exclusively aimed at girls designing in after school game clubs.[38] The 45 games made by 126 girls aged between 11 and 14, turn out to be different from the commercial entertainment games. The games have a cheerfully designed and realistic setting, in which daily social fears and worries are dealt with. Violence is hardly found and if it is, it is of the slapstick type. Contrary to what the researchers expected, the games had a distinct level structure with goals and sub-goals thereby not limiting the play to exploration. The main character was not exclusively female, sometimes ambiguous, and players could often choose their own gender.

Conclusion: the research into game use and preferences generally shows a picture that fits in with the stereotypical Western gender relations. At the same time, this picture does not appear to be static. Changes are expected once girls who started playing games at a young age continue to play as they grow older. They can challenge the industry to develop games not embedded in a macho culture nor confirming the stereotype of the caring woman.[39]

2.4 • EFFECTS

The most important effect of entertainment games on young children is the pleasant experience produced by playing. This also applies to educational games and serious games, although these are mainly aimed at bringing about a learning effect. The use of serious gaming is growing in all sorts of areas in the Netherlands, particularly in primary and secondary education (see Chapter 10 on Education).

Below we will discuss a few specific effects of entertainment games that have been determined by scientific research. The effects are usually unintentional and sometimes negative. We will start with the positive effects.

POSITIVE EFFECTS

playing computer games has, just as traditional games do, a positive social effect on many children. In other words, they learn how to interact with each other. This effect is particularly evident at the end of the primary school and is slightly stronger among boys than girls.[40]

Playing together with peers and brothers or sisters is sometimes based on competition, but it can also be a matter of 'watch and see how others manage'.[41] The nowadays common multiplayer options of many games makes playing together so easy that it is becoming ever more self-evident. This particularly applies to online variations such Runescape and World of Warcraft. Some researchers even find it worrying for the social development if children do not play any games or online games at the start of adolescence.[42]

A second positive effect is of a cognitive nature. There are several types of cognitive effects and they are all strongly dependent on the type of game. A short and quick casual game on the internet mainly trains the fundamental types of processing information, for example, in the visual and attention system. A complex 3D video game demanding hours of concentration to finish also calls for higher cognitive functions such as deductive reasoning, planning and decision making.

Most research on cognitive effects on young children have been limited to the basic functions. In one of these studies, 10- and 11-year-olds played the game Marble Madness in which you have to guide a marble through a 3D maze. Their spatial representation skills improved. This was not the case for children who played word games.[43] The improved spatial representation skills were observed for both boys and girls.[44]

In addition, research into memory effects among children aged between 4 and 7 showed that they could remember pictures found within a game better than pictures that were presented in a more school-like environment.[45]

Also verbal forms of cognition appear to benefit from playing games. Meijs carried out PhD research in the Netherlands among children aged between 5 and 15 who regularly played games.[46] Those who loved strategic and simulation games (such as The Sims, Roller Coaster Tycoon, and Lord of the Rings) performed better at the verbal learning test than children who played other games. This verbal difference could be the result of the games, but it could also be the other way around. In that case, better language skills determine the game preference.

Finally, 'attention' is a basic cognitive function that can be influenced by playing games. Attention is important as the processing of information demands concentration. In a study among children aged between 4 and 6 who were training their attention function in a game environment, it was found that they showed more progress than another group of children undergoing alternative attention training.[47] The researchers consider this to be an important indication that the neurally anchored attention function is sensitive to intervention.

An even more complex cognitive function than 'attention' is 'interpreting information'. This has been studied among 8- and 11-year-old children who had to play an unknown game that they had to describe afterwards. The interpretations made by frequent players of games were better than the interpretations by children who play little. A difference was also found that would be expected on the basis of cognitive development. The 11-year-olds gave an interpretation based on the aims of the game while the 8-year-olds came no further than a general evaluation.[48]

Research on the positive effect on young children is modest but many declare their confidence in the positive applications of games. A few American researchers feel that games can be used to physically activate children and combat overweight. They compared the energy consumption of children aged between 8 and 12 who played a normal game sitting at their screen or played Dance Dance Revolution (DDR), for which they had to use their whole body. The DDR players burned twice as much energy as the screen players. This opens all sorts of possibilities for physically oriented games.[49] In this respect, the great success of physical games on the Nintendo Wii is very encouraging.

Other research shows how games can be used in health care.[50] Young diabetics patients were divided into two groups. The first group was given a normal entertainment game to take home, the second group a game about dealing with diabetes. In the game a diabetics summer camp is overrun by nasty mice and rats. The player has to prevent this and at the same time keep an eye on the blood sugar level of the children, inject insulin and manage nutrition programmes. Both games were regularly played in the six months that followed. Those playing the diabetes game intensified their self-control, talked to others more about their disease and gained more knowledge. As a result, their call for emergency help and additional treatment decreased substantially. Similar effects were also found when playing a game for young asthma patients.[51]

HARMFUL EFFECTS

When the games reached an increasingly larger audience in the 1980s, the concern for possible negative effects on children grew among scientists. As with television (in previous years), people were particularly afraid of the effect of violent games. Research was initially dominated by the 'contagion theory', that assumes that games have a direct influence on the players particularly because games offer a more intense experience than many other media types. This theory is still an important aspect in the public discussion on games. This becomes obvious after brutal violent acts – such as shootings at schools - when journalists and politicians directly say there is a link between the violence committed by the offenders and the violent games they played.

Nowadays, media researchers invariably point out that the link between game content and violence is more complex; if only because millions of others also play violent games and would not hurt a fly.[52] Over the past decades, research has concentrated on exactly determining behavioural effects, for example, by finding out if violent games played by children lead to hitting and kicking of friends or wilfully harming others. This has pushed the contagion theory to the background as it has been determined that children who play games, like other media users, process the information actively. Researchers now take the way in which children can understand the content of the games much more into account.

In so-called 'meta-analyses' the results of dozens of effect studies have been combined. These overviews have shown that there is some relation between playing violent games and aggressive behaviour in the real world.[53] The results, however, should be interpreted with caution as something can be said against the execution of many effect studies. In many cases a brief exposure to game content was used, incomparable to a normal game experience. Besides, the effects were only measured directly after playing while it is necessary to determine whether the effects also remain after some time.[54]

Although most of the research is related to adolescents, there are specific indications that violent games also have an effect on young children. Children aged between 6 and 12 who have played a violent game used more aggressive words afterwards and punished their co-players more severely than those who had played non-violent games.[55] Other researchers found that 4- to 6-year-olds were more aggressive towards their playmates and that 8- to 10-year-olds expressed more aggressive thoughts.[56] Older Dutch research has shown that boys aged between 10 and 14, with a preference for violent games, are less likely to help others than boys without this preference.[57]

Copying the confrontations from games is not the only negative effect that games can have. A broader cognitive effect can also occur. After continued exposure to violence so-called scripts can also be created that are stored in memory and are continually adjusted on the basis of new experiences. The development of an aggressive script on the basis of frequent experiences with violent games can contribute to the fact that the negative emotional reaction to violence levels off and children accept violence more easily in their daily life.[58]

Besides the concerns regarding the effects of violent games there is increasing attention being paid to possible game addiction. The reward structure of many games is comparable to that used in gambling and can result in some players finding it difficult to stop.[59] The social commitment of playing online multiplayer games also encourages players to play longer than they intended.

The boundary between enthusiastically indulging in your hobby and addiction is not easily made. Most researchers only refer to addiction if it has negative effects on other areas, for example, conflicts at home or neglecting school work.[60] We have no figures for game addiction among young children. From a survey of Dutch game players aged between 12 and 18 it does appear that 2 to 9% can be considered addicted.[61]

The possible negative effects of games has been the reason for specific advice on the content of games. In the Netherlands the PEGI classification system (Pan European Game Information) an age rating for games that can be bought or rented, was introduced in 2003. Every game is given an age classification (3, 7, 12, 16, 18) and a contents classification (violence, fear, sex, drugs, bad language, discrimination and gambling).[62] The PEGI ratings intend to inform parents and the players themselves, beforehand, of the game content and to warn for possible harm. Since recently, there is also PEGI online, a system in which games played via the internet are given an age and contents rating.

Besides PEGI, which is particularly intended to give information on the possible harmful effects of games, parents can also consult sites such as Weetwatzegamen.nl (meaning: 'Know what they're gaming'). The government and media industry are attempting to support parents in guiding their playing children through these initiatives. It remains to be seen, however, whether the support will be well received.

Research into parental mediation is, unfortunately, still in its infancy in the Netherlands. Our own research among Dutch parents has shown that parents have clear, balanced views on the playing of games.[63] They recognize both positive and negative aspects in the games their children play. They consider the positive effects, such as improved cognitive skills, to be slightly more important than the negative effects. Besides, a number of parents actively intervene with their children's gaming. For example, they set limits to the time spent on gaming or discuss the content of the games.

The extent to which parents set rules for gaming is less intensive than for watching television, probably because most parents have little affinity with games. Watching television together, for example, is a common way for parents to supervise their children, while playing games together usually only occurs with young children and even then to a limited degree.

The concerns parents have on the playing behaviour of their children is usually related to the amount of time they spend doing it.[64] More than 11% of the parents with children aged between 6 and 9 feel their children spend too much time playing games and for 10- to 12-year-olds the group of concerned parents increases to almost 20%. According to the parents, the playing of games among young children is at the expense of social contact with other children while for older children it is at the expense of doing homework.

2.5 • CONCLUSION

Video and computer games offer young children a unique means of entertainment and have therefore captured a substantial part of the entertainment market. Scientific research has shown that children's game preferences are, to a large degree, linked to their level of cognitive development. On the one hand the varied selection of titles provides for this, but on the other hand it appears that the selection is rather limited.

The market is dominated by games that are more geared towards stereotypical male preferences than those of females. The titles aimed at girls almost always have subjects derived from traditional female roles. Children can use the safe environment of games as a laboratory to explore boundaries and try things out, including scary and shocking activities.[65] The games market should expand their assortment to include games that enable boys and girls to not only experiment with their fantasized identity but to also play around with their own place within traditional gender roles.[66]

Scientific research gives us confidence that gaming can have positive effects on children. For example, on their memory skills but also on a social level. On the other hand, there are also chances of negative effects. Games for which children do not yet have the cognitive or emotional capabilities could result in aggressive thoughts or behaviour. There is also the risk of spending too much time playing games resulting in too little time to play outside or do homework.

The risks can be easily contained if parents gather information on the games by consulting, for example, PEGI and non-profit information portals such as 'Weet Wat Ze Gamen' (know what they're gaming). In addition, there are sufficient indications that setting rules for children on gaming at home bears fruit. The sooner parents and children set rules on the titles that may be played and the amount of time to be spent playing, the better. Parents may sometimes have to compete strongly with the appeal of this unique entertainment medium to uphold the rules. The frustration experienced by young players if and when they have to stop in the middle of a game will definitely be expressed and can lead to conflicts.

The position of parents in relation to their children's gaming shall continue to change over the years as players who started in the 80s will now be getting children themselves. The role of supervisor or forbidding supervisor may then develop into that of co-player, and it cannot be ruled out that parents transfer their own enthusiasm to the next generation.

NOTES

1   NVPI. (2009). *Market information:* The general interactive market. Hilversum: NVPI www.nvpi.nl.

2   ELSPA. (2009). *Video games now more popular than ever*. Entertainment and Leisure Software Publishers Association. Downloaded from http://www.elspa.com/about/pr/;
ESA. (2009). *Industry facts*. Downloaded from http://www.theesa.com/facts/index.asp.

3   ESA. (2009). *Industry facts*. Downloaded from http://www.theesa.com/facts/index.asp.
NVPI. (2009). *Market information:* The general interactive market . Hilversum: NVPI www.nvpi.nl.

4   PEGI. (2009). *What are classifications?* Downloaded www.pegi.info.

5   Nikken, P. (2003a). *Computerspellen in het gezin (computer games at home)*. Hilversum: Nederlands Instituut voor de Classificatie van Audiovisuele Media (Netherlands Institute for the Classification of Audiovisual Media, NICAM).

6   Pijpers, R. & Pardoen, J. (eds.) (2009). *Next level*: Dossier over online spelletjes voor kinderen (Next Level: Dossier on online games for children). The Hague: Stichting Mijn Kind Online (My Child Online Foundation).

7   Pratchett, R. (2005). *Gamers in the UK:* Digital play, digital lifestyles. London: BBC New Media & Technology.

8   Rideout, V.J., Vandewater, E.A., & Wartella, E.A. (2003). *Zero to six: Electronic media in the lives of infants, toddlers, and preschoolers*. Menlo Park, CA.: The Henry J. Kaiser Family Foundation.

9   Roberts, D.F., & Foehr, U.G. (2008). Trends in media use. *The Future of Children, 18*(1), 11-37.

10 Nikken, P. (2003a). *Computerspellen in het gezin (computer games at home)*. Hilversum: Nederlands Instituut voor de Classificatie van Audiovisuele Media (Netherlands Institute for the Classification of Audiovisual Media, NICAM).

11 Van Dorsselaer, S., Zeijl, E., Van den Eeckhout, S., Ter Bogt, T. & Vollebergh, W. (2007). *HBSC 2005:* Gezondheid en welzijn van jongeren in Nederland (Health and well-being of youngsters in the Netherlands). Utrecht: Trimbos-instituut.

12 Pratchett, R. (2005). *Gamers in the UK:* Digital play, digital lifestyles. London: BBC New Media & Technology;
Roberts, D.F., & Foehr, U.G. (2008). Trends in media use. *The Future of Children, 18*(1), 11-37;
Sherry, J.L., Lucas, K., Greenberg, B.S., & Lachlan, K. (2006). Video game uses and gratifications as predictors of use and game preferences. In P. Vorderer & J. Bryant (ed.), *Playing video games:* Motives, responses, and consequences (pp. 213-224). London: Lawrence Erlbaum Associates.

13 Nikken, P. (2003a). *Computerspellen in het gezin* (Computer games at home). Hilversum: Nederlands Instituut voor de Classificatie van Audiovisuele Media (Netherlands Institute for the Classification of Audiovisual Media, NICAM).

14 Sutton-Smith, B. (1997). *The ambiguity of play*. Cambridge, MS: Harvard University Press;
Piaget, J. (1960). *The moral judgment of the child*. New York: Basic Books.

15 Vorderer, P. (2001). It's all entertainment-sure. But what exactly is entertainment? *Poetics, 29*, 247-261.

16 Piaget, J. (1960). *The moral judgment of the child*. New York: Basic Books.

17 Valkenburg, P.M. (2004). *Children's responses to the screen. A media psychological approach*. Mahwah, NJ: Lawrence Erlbaum Associates.

18 Vygotsky, L.S. (1978). *Mind in society*: The development of higher psychological processes. Cambridge, MS: Harvard University Press.

19 Valkenburg, P.M., & Vroone, M. (2004). Developmental changes in infants' and toddlers' attention to television entertainment. *Communication Research, 31*(3), 288-311.

20 Subrahmanyam, K., & Greenfield, P.M. (2009). Designing serious games for children and adolescents: What developmental psychology can teach us. In U. Ritterfeld, M. Cody & P. Vorderer (ed.), *Serious Games*: Mechanisms and Effects. London: Routledge, Taylor & Francis.

21 Blumberg, F.C., & Ismailer, S.S. (2009). What do children learn from playing digital games? In U. Ritterfeld, M. Cody & P. Vorderer (ed.), *Serious Games:* Mechanisms and Effects. London: Routledge, Taylor & Francis.

22 Nikken, P. (2007). *Mediageweld en kinderen* (Media violence and children). Amsterdam: SWP;
Valkenburg, P.M. (2004). *Children's responses to the screen. A media psychological approach*. Mahwah, NJ [etc.]: Lawrence Erlbaum Associates.

23 Valkenburg, P.M. (2004). See note 22.

24 Van der Voort, T.H.A. (1982). *Kinderen en TV-geweld:* Waarneming en beleving (Children and TV violence: perception and experience). Lisse: Swets & Zeitlinger.

25 McDonald, D.G., & Kim, H. (2001). When I die, I feel small: Electronic game characters and the social self. *Journal of Broadcasting & Electronic Media, 45*(2), 241-258.

26 Pijpers, R. & Pardoen, J. (red) (2009). See note 6;
Sherry, J.L., & Dibble, J.L. (2009). The impact of serious games on childhood development. In U. Ritterfeld, M. Cody & P. Vorderer (ed.), *Serious Games:* Mechanisms and Effects. London: Routledge, Taylor & Francis.
Von Salisch, M., Oppl, C., & Kristen, A. (2006). What attracts children? In P. Vorderer & J. Bryant (ed.), *Playing computer games:* Motives, responses, and consequences (pp. 147-164). Mahwah, NJ: Erlbaum.

27 Eagly, A.H., & Koenig, A.M. (2006). Social role theory of sex differences and similarities: Implication for prosocial behavior. In K. Dindia & D.J. Canary (ed.), *Sex differences and similarities in communication, 2nd ed* (pp. 161-177). Mahwah, NJ: Lawrence Erlbaum Associates Publishers.

28 Jansz, J., & Martis, R. (2007). The Lara phenomenon: Powerful female characters in video games. *Sex Roles, 56*(3), 141-148;
Smith, S.L. (2006). Perps, pimps, and provocative clothing: Examining negative content patterns in video games. In P. Vorderer & J. Bryant (ed.), *Playing video games:* Motives, responses, and consequences. (pp. 57-75). London: Lawrence Erlbaum Associates.

29 Vosmeer, M. (2010). *Videogames en gender*: Over spelende meiden, sexy avatars en huiselijkheid op het scherm (Video games and gender: About girls playing, sexy avatars and hominess on the screen). Amsterdam: University of Amsterdam (PhD thesis).

30 Carr, D. (2005). *Contexts, pleasures and preferences:* Girls playing computer games. Paper presented at the Digra International Conference, Vancouver.

31 Kafai, Y. (2008). Gender play in a tween gaming club. In Y.B. Kafai, C. Heeter, J. Denner & J. Sun (ed.), *Beyond Barbie and Mortal Kombat.* New perspectives on gender and gaming (pp. 111-125). Cambridge, MA: The MIT Press.

32 Lucas, K., & Sherry, J.L. (2004). Sex differences in video game play: A communication-based explanation. *Communication Research, 31*(5), 499-523.

33 Ruggiero, T. E. (2000). Uses and gratifications theory in the 21st century. *Mass Communication and Society, 3*(1), 3-37.

34 Sherry, J.L., Lucas, K., Greenberg, B.S., & Lachlan, K. (2006). Video game uses and gratifications as predictors of use and game preferences. In P. Vorderer & J. Bryant (ed.), *Playing video games:* Motives, responses, and consequences (pp. 213-224). London: Lawrence Erlbaum Associates.

35 Sherry, J.L., & Dibble, J.L. (2009). The impact of serious games on childhood development. In U. Ritterfeld, M. Cody & P. Vorderer (ed.), *Serious games:* Mechanisms and effects. London: Routledge, Taylor & Francis.

36 Nikken, P. (2000). Boys, girls and violent video games: The views of Dutch children. In: C. Von Feilitzen & U. Carlsson (ed.). *Children in the new media landscape:* Games pornography perceptions (pp. 93-102). Göteborg: Nordicom.

37 Kafai, Y. (1998). Video game designs by girls and boys: Variability and consistency of gender differences. In J. Cassell & H. Jenkins (ed.), *From Barbie to Mortal Kombat:* Gender and computer games (pp. 90-111). Cambridge, MA: MIT Press.

38 Denner, J., & Campe, K. (2008). What do girls want? What games made by girls can tell us. In Y. B. Kafai, C. Heeter, J. Denner & J. Sun, Y. (ed.), *Beyond Barbie and Mortal Kombat. New perspectives on gender and gaming.* Cambridge, MA: The MIT Press.

39 Jansz, J., & Vosmeer, M. (2009). Girls as serious gamers: Pitfalls and possibilities. In U. Ritterfeld, M. Cody & P. Vorderer (ed.), *Serious games:* Mechanisms and effects. London: Routledge, Taylor & Francis.

40 Von Salisch, M., Oppl, C., & Kristen, A. (2006). What attracts children? In P. Vorderer & J. Bryant (eds.), *Playing computer games:* Motives, responses, and consequences (pp. 147-164). Mahwah, NJ: Erlbaum.

41 Orleans, M., & Laney, M. C. (2000). Children's computer use in the home: Isolation or sociation? *Social Science Computer Review, 18*(1), 56-72.

42 Durkin, K. (2006). Game playing and adolescents' development. In P. Vorderer & J. Bryant (ed.), *Playing video games:* Motives, responses, and consequences (pp. 415-428). Mahwah, NJ: Lawrence Erlbaum Associates.

43 Subrahmanyam, K., Greenfield, P., Kraut, R., & Gross, E. (2001). The impact of computer use on children's and adolescents' development. *Applied Developmental Psychology, 22*, 7-30.

44 De Lisi, R., & Wolford, J. L. (2002). Improving children's mental rotation accuracy with computer game playing. *Journal of Genetic Psychology, 163*, 272-282.

45 Oyen, A. S., & Bebko, J. M. (1996). The effects of computer games and lesson contexts on children's mnemonic strategies. *Journal of Experimental Child Psychology, 62*(2), 173-189.

46 Meijs, C. (2008). Verbal learning in school-aged children and the influence of child-related factors, test related factors, and natural context. Universiteit Maastricht: PhD thesis.

47 Rueda, M., Rothbart, M., McCandliss, B., Saccomanno, L., & Posner, M. (2005).

Training, maturation, and genetic influences on the development of executive attention. *Proceedings of the National Academy of Sciences USA, 102*(41), 14931–14936.

48 Blumberg, F.C. (1998). Developmental differences at play: Children's selective attention and performance in video games. *Journal of Applied Developmental Psychology, 19*(4), 615-624.

49 Lanningham-Foster, L., Jensen, T.B., Foster, R.C., Redmond, A.B., Walker, B.A., Heinz, D., & Levine, J.A. (2007). Energy expenditure of sedentary screen time compared with active screen time for children. *Pediatrics, 118*, 1831-1835.

50 Lieberman, D.A. (2006). What can we learn form playing interactive games? In P. Vorderer & J. Bryant (eds.), *Playing video games:* Motives, responses, and consequences (pp. 379-395). London: Lawrence Erlbaum Associates.

51 Lieberman, D.A. (2006). What can we learn form playing interactive games? In P. Vorderer & J. Bryant (eds.), *Playing video games:* Motives, responses, and consequences (pp. 379-395). London: Lawrence Erlbaum Associates.

52 Nikken, P. (2007). *Mediageweld en kinderen* (Media violence and children). Amsterdam: SWP.

53 Anderson, C.A. (2004). An update on the effects of playing violent video games. *Journal of Adolescence, 27*(1), 113-122. Sherry, J.L. (2001). The effects of violent video games on aggression: A meta-analysis. *Human Communication Research, 27*(3), 409-431.

54 Goldstein, J. (2005). Violent video games. In J. Raessens & J. Goldstein (ed.), *Handbook of computer game studies* (pp. 341-359). Cambridge, MA: The MIT Press.

55 Anderson, C., Gentile, D., & Buckley, K. (2007). *Violent video game effects on children and adolescents*: Theory, research, and public policy. Oxford University Press.

56 Lee, K.M., & Peng, W. (2006). What do we know about social and psychological effects of computer games? A comprehensive review of the current literature. In P. Vorderer & J. Bryant (ed.), *Playing video games:* Motives, responses, and consequences (pp. 327-345). Mahwah, NJ: Lawrence Erlbaum Associates.

57 Wiegman, O., & Van Schie, E. (1998). Video game playing and its relations with aggressive and prosocial behaviour. *British Journal of Social Psychology, 37*, 367-378.

58 Huesmann, L. (1998). The role of social information processing and cognitive schema in the acquisition and maintenance of habitual aggressive behavior. In R. Geen & E. Donnerstein (ed.), *Human aggression:* Theories, research, and implications for social policy (pp. 73-109). San Diego, CA: Academic Press.

59 Griffiths, M. (2007). Videogame addiction: fact or fiction. In T. Willoughby & E. Wood (eds.), *Children's learning in a digital world* (pp. 85-103). Oxford: Blackwell.

60 Griffiths, M. (2007). Videogame addiction: fact or fiction. In T. Willoughby & E. Wood (eds.), *Children's learning in a digital world* (pp. 85-103). Oxford: Blackwell.

61 Lemmens, J.S., Valkenburg, P.M., & Peter, J. (2009). Development and validation of a game addiction scale for adolescents. *Media Psychology, 12*(1), 77-96.

62 PEGI. (2009). *What are classifications?* Downloaded from www.pegi.info

63 Nikken, P., & Jansz, J. (2006). Parental mediation of children's videogame playing: A comparison of the reports by parents and children. *Learning, Media, & Technology, 31*(2), 181-202.

64 Nikken, P. (2003b). Ouderlijke zorgen over het 'gamen' van hun kinderen

(The concerns of parents about their children's gaming). *Pedagogiek, 23*(4), 303-317.

65 Jansz, J. (2005). The emotional appeal of violent video games for adolescent males. *Communication Theory, 15*, 219-241.

66 Jansz, J., & Vosmeer, M. (2009). Girls as serious gamers: Pitfalls and possibilities. In U. Ritterfeld, M. Cody & P. Vorderer (ed.), *Serious Games: Mechanisms and Effects.* London: Routledge, Taylor & Francis.

"Saturday is computer day"

Brothers Erasmus (6) and Rembrandt (8) from Amsterdam enjoy playing online games. Saturday is traditionally their computer day. They play together, each on their own laptop. "And when our mum is travelling we're allowed to play a bit more often."

Erasmus: "I usually play via Speeleiland.nl or via Spele.nl. I enjoy games in which you have to discover things. Such as Spiderman, or Batman. I like these games because you have to go somewhere and you don't know where and then at the end you have to fight. No, I never find that scary. But I don't like all games. I don't like racing games or girls games. My four-year-old sister always plays My Little Pony or dress-up games. I've also tried them but didn't really enjoy them."

Rembrandt: "I often go to Gamezhero.nl. I usually play a game with monsters who shoot at flowers. You have to kill all the monsters. And if you've managed that, you go to the next level and it becomes more difficult. Actually, I like all games on Gamezhero. But I also play *Age of Empires* a lot. That's a CD-ROM game in which you're a colonist. It's situated in a time when guns were only just invented. I like strategic games, they are often not on the internet."

Erasmus: "I often play on my Playstation, I like this the most. But the nice thing about playing on the internet is that the games are free. There are games for which you have to pay, but I never play those."

Rembrandt: "On Saturdays we're always allowed on the computer, that's our computer day. We also have a TV day, a friends day and toy days. But if my mum is away, travelling or so, my dad allows us to use the computer more often. My mum is not too keen on it, but my dad is. Making websites is his job, so he is also at the computer a lot."

Erasmus: "We have a lot of computers at home. And we also all have our own laptop. If my brother is behind his laptop, I also often want to use mine. Then I play together with my friends Milan and Rachid, for example."

Rembrandt: "We know how to use the the internet. I never fill in my details anywhere, because then someone can hack your computer. That has already happened to me once – I saw the screen change and all my details were gone. I understand everything. Only if I want to install something my dad helps me."

Kourosh Ajamlou, Erasmus and Rembrandt's father: "Haha, yes, that's right, they are allowed to use their laptops when their mother is away. I do keep an eye on them, I never let them play alone. Rembrandt is very computer-minded, he knows everything. Erasmus follows him. Rembrandt often plays strategic games that require analytic thinking. Like *Age of Empires*. That game is really complicated and I myself wouldn't be able to play it. I buy these games for him, he gets bored quickly of normal games."

"The online games are all more or less the same. Only the characters are different, and they have very little content. There are also many bloody and violent games which I don't let them play."

"We recently watched a documentary about children addicted to games, they really became zombies. Sohave seen now that they can easily get addicted to the computer. But still…. if it were up to them, they would play games day and night, which is why you have to set rules."

3

Casual games

Nathalie Korsman
Menno Deen

Nearly all Dutch children aged between 8 and 12 play casual games: short online games on game portals such as Spele.nl, Funnygames.nl and Speeleiland.nl like with playing the Game Boy (kids call this 'DS'ing'). They often play these games alone, but just as with Game Boys (now DS) you often see children sitting together in front of the screen. The portals themselves can also contain social elements.

Casual game portals respond to the tendency of social interaction also seen in 'larger' games such as Runescape, and social networks such as Hyves. Think, for example, of the possibility to play with other visitors, comment on games, or chat while playing (especially with multiplayer games).

Playing casual games can be fun and educational but it can also have risks. The games can, for example, be violent and the social interaction can sometimes go off the rails. This can include sexually tinged remarks, bullying, bad language and theft or virtual theft.

In this chapter we will discuss the following:
• The casual game concept;
• Game portals;
• Quality of the games;

- Respond to needs;
- Social tendencies;
- Learning results;
- Risks and recommendations.

3.1 • THE CASUAL GAME CONCEPT

At the start of this chapter we mentioned – to define the scope – that casual games are online games (browser games) but the scope of quickly playable games is not limited by that definition. They can also be found offline (such as Solitaire and Minesweeper on Windows-PCs), on mobile phones (Java games), and on the shelves of specialist game shops[1] and, they are not always free of charge. Think of, for example, the waitress game *Delicious – Emily's Taste of Fame* (€ 30,-), the pinball machine game *Peggle* (€ 19,-) or the shooting game *Geometry Wars: Retro Evolved* (€ 5,-).[2]

We use the following two criteria to differentiate casual games from other games:
- **short playing sessions** (± 5 minutes per level), in other words: limited time to attain a goal. For example, emptying a field of jewels within a certain time in the game *Bejeweled* or preventing a full playing field in *Tetris*[3] ;
- **steep learning curve**, in the original definition (from cognitive psychology) of 'getting the hang of something quickly'. The gameplay is easy and the rules are simple. If it involves levels, they gradually become more difficult and the player discovers new strategies and ways of playing. The industry uses the slogan *easy to learn, hard to master* for this.

A game that complies with the above criteria is the arcade classic *Pong*.[4] Each game lasts a few minutes (short playing session), the gameplay is easy and the rules are simple (steep learning curve). *Pong* was successful in the arcades of the 70s. They were filled with these types of arcade games that could be played for twenty five cents. The modern game portals can be seen as the online version of those classic arcades.

Another similarity between arcade games and casual games is the so-called *coin drop*. Successful arcade games were originally designed to find the perfect balance between playing time on the one hand and playing costs on the other. This is the 'coin drop' principle. The game should not be too short as players feel taken in but they should not last too long either as they then play too long using just one coin, not bringing enough revenue. Casual games build on this 'coin drop' principle, although the earnings models nowadays are more complex than just inserting a coin.

Game portals have taken the place of the original arcades. There are more than a dozen successful online games portals in The Netherlands. The four most important and popular ones are listed in this paragraph.

Spele.nl contains about 5000 games and is by far the most popular site for casual games. This website is visited by approximately 10 million people every month. About 30% of these visitors is aged between 6 and 12, according to Spele.nl's information for its advertisers. This means almost all Dutch children. Figures from My Child Online Foundation's dossier *Next Level* also confirms this site to be very popular. 74% of all 8- to 12-year-olds visit Spele.nl and as many as 80% of 12-year-olds regularly play a game here.[5] The site is a product of the internet company I-Med Internet Services. I-Med has four employees. The founders were mere teenagers when they started Spele.nl.

Funnygames.nl lists around 3000 games is owned by Tibaco Internet Media, Eindhoven, The Netherlands, since 2002. As was the case for Spele.nl, the hobby of two teenagers resulted in Funnygames.nl. At the beginning of 2009, the portal attracted approximately 3 million visitors per month.

Spelle.nl has more than 5000 games and is published by Plox Internet Media. This site too started out as a hobby, to earn money from banners show on the portal. The site now has approximately 3 million visitors per month. One-third of these visitors is under the age of 13, according to Plox.

Spelletjes.nl claims that it attracts 1.8 million visitors per month. The 200 employees of owner Spil Games supplies games to more than 50 sites in more than 20 languages. The company reportedly reaches more than 100 million unique visitors with its websites worldwide. Its headquarters are based in Hilversum, the Netherlands and they have offices in Poland and China.

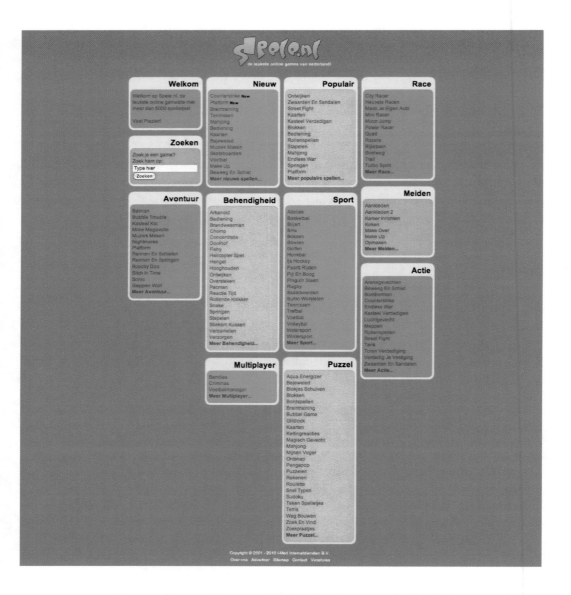

Figure 3.1 • Game portals have taken the place of arcade games. Spele.nl is by far the most popular Dutch game portal.

Most casual games offered online are of poor quality. On game portals you mainly find games made by hobbyists, freelancers and advertising producers:

- hobbyists often give existing games a facelift (often bloody or politically incorrect);
- freelancers (also known as 'indies', short for 'independent game designers') like to try out new game principles;
- advertising producers make games to promote products.

Furthermore there are professional game studios, such as *Zylom Studios* and *PopCap Games*. As these companies want to reach as many people as possible, their games are seldom bloody or of a sexual nature. They generally produce high quality games, beyond all criticism. But such games – like the popular *Bejeweled* – are then copied, revised and offered to game portals once more.

3.4 • RESPOND TO NEEDS

Casual game portals draw many children. This could be attributed to the steep learning curve of casual games mentioned earlier, the extensive range, The increasingly social nature of the playing environment. Portal managers and game developers, however, also make use – consciously or unconsciously – of the human need to feel competent, autonomous and connected to others.[6]

COMPETENCE AND SCAFFOLDING

Children who play casual games often say: "I almost made it! Next time I'll get him!" And they are not wrong. The steep learning curve of a casual game enables a child to actually be able to go a stage further each time. This can give them a feeling of competence.

In education, this is called *scaffolding*: gradually offering new information (gameplay elements) in line with the level of knowledge and social and cultural environment of the pupil (in this case the playing child).[7] Good casual games 'scaffold' the progress of their game in such a way that the goals of the game are within reach of the child. In addition, the games communicate the progress made.

Think of *Pac-Man*.[8] This game suggests that you can eat all the pills in the playing field: an apparently attainable goal. And it works for a long time: pills eaten, sub goals and goals attained, rewards received. Even the poorest player can easily attain a goal, giving a feeling of competence. This feeling increases the appeal of games.[9]

AUTONOMOUS

The freedom a child has to make their own choices – in type of game, strategy and playing type – increases the feeling of being autonomous and thereby the motivation to play. In other words: the more you yourself can determine the agenda, the more appealing the situation becomes.

It is for this reason the game portals are so appealing. They offer a banquet of games: different kinds of games, with different strategies and different playing styles. Children can create their own challenges and solve them in their own way.

Further information:
- Nick Yee has carried out extensive research on the playing motivations and playing styles of large online games.[10]
- Alessandro Canossa has derived various playing styles from the series *Hitman*.[11]

CONNECTION

The third basic need that can be motivating is the possible fulfilment of relationship: feeling connected with others. This social need can also be clearly seen in the arcade games mentioned earlier. They function as a hang-out for young people, with their own social roles and rules, a language of their own, their own codes of conduct and sign language.[12]

It is precisely in the environments where players can see each other or communicate with each other (for example, game arcades, but also online multiplayer games, game related websites and *sneak previews*) that the social value of casual games is most clearly seen.

3.5 • SOCIAL TENDENCIES

Social behaviour and related social rules can be found in online social netwerks (such as Hyves and Facebook) and in multiplayer games (such as Runescape and World of Warcraft). In the case of casual games it is slightly more complex.

Media researcher Marinka Copier describes the interaction between 'real' social rules and online behaviour. This results in new social spaces in online games and social networks, with their own standards and values. People meet each other, connect to each other, or exclude others. Because almost all children play casual games, it is not surprising that their 'real' social networks (in real life) are also formed around casual games and in relation to casual games.

These new networks have their own social model. Communications specialist Mia Consalvo describes how some models have a hierarchic structure, based on game capital. Game capital is the knowledge of and the skill in games. Children with a large game capital are at the top of the social ladder, while newbies (starters) can be found at the bottom. Children with large game capital steal the show on YouTube, with clips on tour de forces in games and the accompanying top scores. The manager of the YouTube channel *Nintendodsg,* for example, shows how the classic Super Mario Bros can be finished very quickly.

CURRENT STATE OF AFFAIRS

The social interaction that takes place on network sites and online multiplayer sites is currently not or hardly found on the casual game portals popular among children. Apart from the occasional chat or response possibilities the most important social interaction consists of the ratings system that users can use to make their appreciation for games known. The game with the most votes moves to the top of the list. This results in a collective decision on good games.

The choice for an assessment system is not surprising in view of the greatly variable quality of the casual games on offer. This feature for rating games was perhaps not really developed to stimulate social interaction but it seems to be changing the platforms nonetheless.

NEW DEVELOPMENTS

The developers of casual games appear to be re-discovering the social value of games. For example, social networks (Dutch social network Hyves and its international opponent Facebook) show an increasing number of casual games on offer. Advertisers are also making use of this trend. Their 'advergames' (blending the words advertising and gaming) are not developed by Hyves or Facebook themselves but can be connected to the social network by external developers. The trend is to use the profile information of players and their competitors, such as photos, age, birthday and address details, in these games.

An example of this type of casual game is *SkyDiver*, developed for Dutch social network Hyves.nl This game can be placed on a Hyves page as a 'gadget (Hyves' word for embedded external widgets, games and movies)'. Players can invite friends to play via the Hyves network. According to various sources SkyDiver has been played more than one million times and can be found on more than 50,000 profile pages.

Similarly, in the virtual worlds a growth in casual games has been shown. Some worlds are entirely dedicated to short playing sessions and are therefore called casual online multiplayer games. Examples of popular casual online multiplayer games, especially for children, are *Club Penguin* and the recently launched *Free Realms*. Casual game portals also follow the social example of Hyves and online multiplayer games. Players have to register increasingly often to be able to chat, give comments, dress a virtual character, use game attributes or currency, or to be able to play at all (for example, Doof.com or Habbo.nl).

Figure 3.2 • 'Hyves' users could put the game Skydiver, developed on behalf of Centraal Beheer Achmea, on their own page.

3.6 • LEARNING EFFECTS

Children can develop various skills by playing casual games.[13] These skills can be roughly divided into three categories:
• Media-specific skills;
• Metacognitive skills;
• Social skills.

MEDIA-SPECIFIC SKILLS

Participating in casual game portals teaches children to deal with digital and interactive media.[14] It teaches children how games are played and how they can understand the rules.[15]

A casual game, however, is not just *played*, it is also *experienced* in a number of ways. Children visit game websites, chat and post notes on casual games. Many children make their own 'fan websites', and some children are even capable of making their own games, thereby also developing media production skills.

METACOGNITIVE SKILLS

Playing casual games not only trains children in the playing of the games themselves, including all the rules and skills that apply to a specific game, but it also trains them in various metacognitive skills, such as the ability to solve problems, context knowledge, learning by trial and error and 'not being afraid to make mistakes'.[16]

With regard to the context knowledge: many casual games confront children with a problem. The child then has to think of a solution to the problem. Because games always describe a certain space within which the rules of the game apply, the game creates a context, where 'knowledge' has a value. A good player will understand in which context a particular piece of knowledge is required.

Another quality of casual games is that children learn to progress in the game by trial and error and are not afraid to make mistakes. Games are therefore sometimes compared to a sandbox, or 'a safe sandbox to learn in'.[17] In this way, making mistakes looses its negative connotation such as 'personal failure' and 'stupidity', and – turned around – are as such seen as an essential part of the learning process.

It has, to an increasing degree, become expected of children that they work together or make strategic decisions in a team. Playing games can help them to achieve this as games – as they become more social – require an increasing amount of negotiation. Players also learn to project themselves in social hierarchies. In addition, children can and dare to experiment in games with various roles or characters.[18] They play with their identity and that of others.

3.7 • RISKS AND RECOMMENDATIONS

The risks of harmful content – for example, violent or bloody games, misleading advertisements, etc. – have already been discussed in detail. See, among other, the document 'Next Level' (Stichting Mijn Kind Online, 2009). Below we will focus on the social risks.

Current casual game portals are under extensive development and increasingly look like the arcade games of the past. Children do not just sit in front of their PC but will increasingly come into contact with other players. As was the case for arcade games, new social standards and manners will be created for online portals. And as in the past, online meeting places could be associated with child molesters, bullying and fraudulent practices. This is a pity, as children can also learn a lot on casual game portals (see above).

There is, however, an important difference between arcade games on the one hand and casual game portals on the other. Namely, the fact that casual game portals, in principle, make it possible to either just play with friends or to come into contact with strangers. Not all portals actually have the option of just playing with friends and acquaintances, but it is, as such, possible. We would therefore like to recommend that, as is the case elsewhere on the internet, an explicit distinction is made between open networks and closed networks for casual game portals.

OPEN AND CLOSED NETWORKS

In an open network everyone is connected with everyone, including strangers. Examples of open networks are World of Warcraft and the advergames on Dutch social network Hyves. These spaces are designed as public meeting areas where the operators try to couple as many players to each other as possible.

In a closed network the only connection is with friends and acquaint-ances. An example of a closed network is MSN Messenger. Children can chat with someone if they accept an invitation. The children then gen-erally know the people they talk to and the chances of undesirable con-tacts decrease.

Well-known casual game portals like Spele.nl and King.com do not yet have the possibility to just play with 'real friends' in a protected envi-ronment. But on some other portals children can play against each other anonymously. The nickname of the competitor is then visible but it is not possible to chat or see the profile page. An example is Well-games.com.

ADDITION TO PEGI

Being that many casual game portals are, as such, open networks, is usually not communicated well. This results in children not being suf-ficiently prepared for unfriendly comments or improper requests by strangers. There is room for improvement in the current classification system used by PEGI.

PEGI's classification system is currently limited to the basic 'online' la-bel if an offline game has an online component. As such this is problem-atic as the PEGI online quality label, that provides age information to parents, is only given to sites for which the entire content has been clas-sified (on the basis of age, violence, sex, etc). For large game producers like EA-Games (producer of, among other, The Sims) this is not really a problem but for game portals such as Spele.nl, with thousands of un-classified games, it is. This is why the 'Next Level' editors (Stichting Mijn Kind Online, 2009) recommends operators to differentiate their range of games by suitability for certain ages: all games for particular age groups together. And preferably, of course, a PEGI classification for each individual game.

This recommendation is, however, only directed at the content. The social context should also be clearly indicated. It is therefore recom-mended that the current icons are supplemented with icons for open networks and closed networks:
• the icon 'open network' will then mean: an online community where children may come in to contact with strangers;
• the icon 'closed network' indicates that players can only play with each other if 'invited'.

Casual game portals offer a reasonably safe place to play together. It is very likely that children further develop their media-specific, metacognitive and social and cultural skills in these online environments. It is therefore important that these portals continue to guarantee this safety. The ever increasing social content of playing games is putting their to the test.

The bad reputation of the arcade games could also affect casual game portals. In order to avoid this, children and their carers should be informed about the network they are connecting to. A supplement to the PEGI information with icons for open and closed networks could contribute to this.

NOTES

1 Bennallack, O. (2008). Casual Biz Models No. 1 – Retail Distribution. *Casualgaming.biz*. Obtained 23 September 2009, from http://www.casualgaming.biz/news/27459/casual-biz-models-no1-retail-distribution.

2 Zylom Studios. (2009). *Delicious - Emily's Taste Of Fame*. Zylom;
PopCap Games. (2007). *Peggle*. PopCap Games;
Bizarre Creations. (2003a). *Geometry Wars: Retro Evolved*. Microsoft Game Studios.

3 Zylom Studios. (2009). See note noot 2.
PopCap Games. (2001). *Bejeweled*. PopCap Games.
Pajitnov, A., & Gerasimov, V. (1984). *Tetris*. Nintendo.

4 Atari Inc. (1972). *Pong*. Atari Inc

5 Pijpers, R. & Pardoen, J. (red). (2009). *Next Level: Dossier over online spelletjes voor kinderen* (Next Level: Dossier on online games for children). Mijn Kind Online. Obtained from www.mijnkindonline.nl on 10 September 2009.

6 Ryan, R.M., & Deci, E.L. (2000). Self-determination theory and the facilitation of intrinsic motivation, social development, and well-being. *American psychologist*, 55(1), 68–78.

7 Verenikina, I. (2003). Understanding scaffolding and the ZPD in educational research. In *Australian Association of Educational Research Conference, Auckland, New Zealand*.

8 Namco. (1980). *Pac-Man*. Namco.

9 Bandura, A. (1997). *Self-Efficacy in Changing Societies*. Cambridge University Press;
Ryan, R. M., & Deci, E. L. (2000). Self-determination theory and the facilitation of intrinsic motivation, social development, and well-being. *American psychologist*, 55(1), 68–78.

10 Yee, N. (2006a). Motivations for play in online games. *CyberPsychology & Behavior*, 9(6), 772–775;
Yee, N. (2006b). The demographics, motivations, and derived experiences of users

of massively multi-user online graphical environments. *PRESENCE: Teleoperators and Virtual Environments*, *15*(3), 309–329.

11 IO Interactive. (2007). *Hitman (Series)*. Eidos Interactive;
Canossa, A. (2005). Designing Levels for Enhanced Player Experience Cognitive tools for gameworld designers. IO Interactive / Denmark's School of Design;
Canossa, A. (2007). Weaving Experiences Values Modes Styels and Personas. IO Interactive / Denmark's School of Design;
Canossa, A. (2008). Towards a theory of Player: Designing for Experience. IO Interactive / Denmark's School of Design;
Canossa, A., & Drachen, A. (2009). Play-Personas: Behaviours and Belief Systems in User-Centred Game Design.

12 Fisher, S. (1995). The amusement arcade as a social space for adolescents: an empirical study. *Journal of Adolescence*, *18*(1), 71-86;
Ashcraft, B. (2009). Arcade Mania: The Turbo-charged World of Japan's Game Centers. Kodansha International.

13 Egenfeldt-Nielsen, S. (2006). Overview of research on the educational use of video games. *Digital kompetanse*, *1*(3), 184-213;

Egenfeldt-Nielsen, S. (2005). Beyond Edutainment: Exploring the Educational Potential of Computer Games;
Kebritchi, M., & Hirumi, A. (2008). Examining the pedagogical foundations of modern educational computer games. *Computers & Education*, *51*(4), 1729-1743;
Van Eck, R. (2006). Digital game-based learning: It's not just the digital natives who are restless. *Educause Review*, *41*(2), 16.

14 Gee, J.P. (2003). What Video Games Have to Teach Us About Learning and Literacy (1e ed.). Palgrave Macmillan;
Prensky, M. (2006). *Don't Bother Me Mom--I'm Learning!* Paragon House Publishers.

15 Bogost, I. (2007). Persuasive Games: The Expressive Power of Videogames. The MIT Press.

16 Deen, M. (2007). Versnelde Kennisontwikkeling in Games. Utrecht University;
Shaffer, D.W. (2008). *How Computer Games Help Children Learn*. Palgrave Macmillan.

17 Sihvonen, T. (2009). Players Unleashed! Modding the Sims and the Culture of Gaming. University of Turku.

18 Copier, M. (2007). Beyond the magic circle: A network perspective on role-play in online games. Utrecht University. zie 22.

"In Habbo you can remove bullies with one click"

Ross and Julian (both 12) are best friends. They not only talk to each other 'live' and on the phone, they also meet in Habbo Hotel, a virtual world that is all about making and keeping social contacts. Both boys have an autistic disorder and go to a special school.

Ross: "On Habbo you can hang around with others, that's fun. If Julian's online, I often go to him. Sometimes we talk on the phone at the same time. We tell each other things that we would otherwise forget. For example, if my mum hits her head, I can tell him that straight away."

Julian: "I have fun and think up jokes with children I know from school. About mistakes that we find on Habbo, for example. You can turn around a fireplace and stand in the flames. We laugh about that."

Ross: "I often go to the club room 'We hate Scammers'. Scammers are Habbos who try to steal things from you. They ask you for your password."

Julian: "Or they try to steal your furni [popular things on Habbo for which you have had to pay real money, ed.] by asking you out. I once went out with a girl on Habbo, but that didn't work out, so I broke it off."

Ross: "Someone also asked me out but that was only to make someone else jealous. So I didn't accept."

Julian: "It is sometimes dangerous to talk to people you don't know because they can bully you. Someone also called me names. Fortunately, you can then remove that person. Easier than in real life. But as in real life you can also become insecure on Habbo, if you see bullies."

Ross: "I therefore prefer to keep the fact that I have Asperger to myself. They can bully you because of this. In real life everyone knows. That is fine, because then they know how to deal with me. I don't know whether there are more Habbos who are autistic, but I'm sure there are."

Julian: "It's difficult sometimes that you don't know. For example, if someone suffers from ADHD, you know you have to take it easy. I once said something wrong, as a joke, and that person became angry. He said: "stop it, I'm going to kill you." You throw someone like that off your friend list."

Ross: "This is sometimes frightening but Habbo is usually just fun to do and it's difficult to stop. I once did the Elfstedentocht (Habbo's online version of the traditional skating competition between 11 cities in the Netherlands) and had to visit 200 rooms. I reached room 100 and got stuck. I had to cry just like those children that are addicted. Now my mother doesn't allow me to spend so much time on it and things have improved."

Simone, Ross' mother: "Habbo is great for Ross and Julian as there are no faces and appearances and no external stimuli. They are in control. Exciting things do happen of course, as with playing games, but they are in control. Although it can go wrong sometimes. This would sometimes make Ross angry and upset, especially when he was younger. He would, for example give his password to a friend and she would steal all his things. For autistic children things like this are worse because at a time like this their world collapses. He would be terribly sad all day and not want to do anything because something had gone wrong. I did away with Habbo for a while. But he can now cope better with it."

4

Everyday creativity
in virtual worlds

Mijke Slot

-j-a-n-n-e-s-: ":O, my coke is finished :(brb, getting new one"

Jannes gets his Habbo character to walk to the soda machine in the corner of my virtual interview room. By pressing a button, his avatar takes a bottle of soft drink from the machine. Satisfied he walks back to his spot on the sofa and takes automatically generated sips. His thirst has been quenched.[1]

Virtual worlds could be defined as seemingly existing worlds. We perceive them but they do not physically exist. This chapter deals with creativity in virtual worlds generated by computers and to be visited via an online connection.[2] Users in such worlds are often represented by avatars (virtual characters), that can communicate with other characters within the virtual world.[3] Well-known virtual worlds include: Second Life, World of Warcraft, Little Big Planet and Habbo (previously: Habbo Hotel).

The dividing line between virtual worlds, social networks and online (multiplayer) games is not always equally clear. Some virtual worlds incorporate playing elements, to a greater or lesser extent. An example of a virtual world in which the playing element is the key element is World of Warcraft. In addition, some virtual worlds function as social networks. Users can then maintain a profile, meet new people and communicate with each other.

A description of virtual worlds for children is given below. We have focused on what creativity means for the use of a virtual world. The information was obtained from a study carried out among visitors the Dutch version of virtual world Habbo (www.habbo.nl).

4.2 • VIRTUAL WORLDS FOR CHILDREN

Over the past years, various new virtual worlds for children have appeared.[4] Mainly commercial parties appear to focus on this market, including traditional offline parties such as Mattel and Disney, but also new online producers such as Sulake, Young Internet Inc and watAgame. These worlds are popular. A survey carried out in 2008 shows that almost three quarters of British children aged between 7 and 12 have at some point visited a virtual world.[5]

TARGET GROUPS AND DESIGN

One of the first virtual worlds for children was Whyville (www.whyville.com). Whyville was created by an American professor, launched in 1999 and is aimed at children aged between 8 and 15. The world of Whyville is of a mainly educational nature. Whyville's appearance certainly gives away the fact that it is one of the oldest virtual games. The avatars in Whyville are not complete figures but merely 2D torsos, consisting of a head and half a body without arms and legs.

Over the years, particularly from 2004-2005 on, an increasing number of worlds have been added, directed at very young children. These are often 2D worlds, designed in cartoon style with animal avatars. For example:
• Panfu (www.panfu.nl) is an online 2D environment for children from the age of 4, with panda avatars. Panfu was developed in 2007 by the German company Young Internet Inc. The Dutch version has been available since 2008;
• Bollykids (www.bollykids.com), which was also developed by Young Internet Inc., is a virtual world for young cildren aged between 4 and 10. This world is currently only available in German;
• Club Penguin (www.clubpenguin.com) is also directed at a young target audience. It is an English language virtual world for children aged between 6 and 14, with penguin avatars. Club Penguin was launched in 2005 and was bought by Disney in 2007;

• Shidonni (www.shidonni.com) is an Israeli initiative that started in 2009. Children can draw their own animal, that is then brought to life and takes them into the virtual world.

In addition, there are virtual worlds directed at slightly older children, from the age of about 8 or 10. These worlds are also usually 2D and a number are specifically aimed at girls. Examples of these worlds for slightly older children are:

• Barbiegirls (www.barbiegirls.com) by Mattel is a virtual world for girls, in which you choose a Barbie as avatar;
• goSupermodel (available in Dutch at www.gosupermodel.nl) designed by the Danish watAgame is also meant for girls. In this case the avatars look like models.
• Habbo (available in Dutch at www.habbo.nl), previously Habbo Hotel, is a very popular virtual world for a mixed target group, consisting of children and teenagers aged between 10 and 20. The average age in the Dutch community is 14. The 3D world was created in 2000 by the Finnish company Sulake. Sulake now distributes Habbo in 31 countries. Avatars in Habbo look like pixel dolls and are called Habbos.[6]

Figure 4.1 • Barbiegirls.com, a virtual world for girls.

- Taatu (www.taatu.com) is a 3D virtual world created in 2005 by the Belgian company Taatu and is suitable for teenagers and youngsters aged between 10 and 19. A Dutch version has been available since 2006.[7]

ACTIVITIES

Virtual worlds try to increase the involvement of their residents in all sorts of ways. In Club Penguin children can, for example, play games with other participants. They can make friends and chat with other penguins. This is also possible on Panfu but they can also choose to play various adventures. Children who visit Taatu (and are then called Taatus) can create their own character, dress themselves and decorate their loft, make friends, chat and enter competitions. In this respect, Taatu is very similar to Habbo. Designing your own avatar and furnishing your own personal space is often limited to the possibilities created by those offering the service. The main aim of the games is entertainment. Players can, for example, jump from a diving board in Habbo and catch fish in Panfu.

The only world that differs from the others in this respect is Whyville. In Whyville, children (called Whyvillians) can take part in all kinds of activities, as in the other worlds. Whyville's approach, however, is mainly educational. The environments and games are sponsored and developed by a large number of American government organisations, universities, non-profit institutes and companies. For example, NASA (called WASA in Whyville) has a laboratory where Whyvillians can experience weightlessness, and the Field Museum in Chicago has built a beach where children can learn about coral reefs. Besides playing educational games, children in Whyville can set up their own business and design parts for avatars (and sell them). In addition, they can also write articles for the Whyville Times - a weekly newspaper.[8]

COMMERCE

Almost all virtual worlds can be entered free of charge but almost all of them charge for extra possibilities. This is not surprising considering that most worlds are provided by commercial companies and the costs of allowing many children to participate in a virtual world at the same time are high.

Club Penguin participants can, for example, buy items, such as clothing for their penguin, furniture for their igloo or pets named Puffles – if they become member. Membership costs €4,95 per month. You can also become a member of Panfu, in this case for €3,90 per month, thereby suppressing, among other, the advertisements. In order to be able to log on to Bollykids you first have to buy a Bolly, an actual cuddly toy, for €14,95. The code attached to this cuddly toy will permit children to log on and play games for six months. In doing so they can collect points to care for their virtual Bolly cuddly toy.

GoSupermodel participants can earn goMoney playing games so as to decorate their room and buy clothes. Players can also buy goMoney by means of a subscription (€10,- per month or €25,- for 3 months).

In Habbo, furniture (called 'furni'), playing certain games and acquiring a premium subscription (Habbo club membership) cost extra money. Most furnis in the Dutch community 3 to 6 credits, approximately 45 to 90 euro cents.[9] At around €3,- per month Habbos can become member of the Habbo Club. Habbo Club members (HC members) have a greater choice of rooms, furni and clothing. HC members also receive a special HC piece of furniture free of charge each month. Credits can be obtained in several ways. For example, by calling a service number, text messaging, or with online payment systems such as iDeal, MinTix or Wallie. A monthly limit of approximately €16,- has been set on most payment methods.

SAFETY

In all virtual worlds for children much attention is paid to safety. The children are informed of the rules for participation before they enter the world. They are usually not, for example, allowed to exchange personal details such as telephone numbers and address. In some worlds the children are advised not to use their real names. They also need parental consent under a certain age.

Many worlds offer their residents the possibility to call upon a moderator when necessary, for example, if other residents do not adhere to the rules. A large M becomes visible on the screen, for example, in Panfu, that can be used to call upon a moderator.

Chatting is another type if interaction in which it is also attempted to guarantee the safety of the participants. Some worlds, such as, Barbiegirls, only permit chatting predefined sentences. In other worlds, you first need to obtain a chat diploma before being allowed to chat. Habbo uses a word filter that automatically replaces 'wrong words', such as swear words, by the word BOBBA. Moderators can also follow conversations live.

4.3 • CREATIVITY IN VIRTUAL WORLDS

Some researchers claim that creativity cannot exist in (commercial virtual) worlds.[10] Others, however, say that virtual worlds derive their right to exist from the creativity of their residents. This calls for further analysis. But how can creativity be analysed in virtual worlds?

A definition often used for creativity is: the ability to produce work that is new (original, unexpected), of high quality and suitable (useful, meet the requirements of a certain task).[11] But this definition does not give a definite answer on creativity. Questions remain when using this definition, for example, about how new or suitable a product must be and how quality is determined so that the product can be given the creative label.[12] To this can be added the fact that creativity is not only expressed in products themselves but can also be used in, for example, processes, ideas or activities.

Figure 4.2 • Screenshot of Habbo. By showing own initiative or by doing unexpected things, children can be creative with their avatar in a virtual world.

When people talk about creativity, and creative people, they are usually referring to artists, writers and inventors. These people possess an exceptional amount of creativity and can spend their whole life trying to achieve recognition for this creativity in a particular area (for example, as painter). Being recognized and communicating this recognition are therefore elements frequently mentioned in combination with creativity. It is often psychologists who determine which factors and characteristics influence this creativity.[13] Creativity is, however, not limited to for artists, writers and inventors. Everyone is creative to some degree. Children are still involved in artistic activities such as drawing, acting and playing music.[14] They also express their creativeness in virtual environments.

In this chapter we have chosen for a pragmatic approach and study everyday creativity in the domain of virtual worlds. In which way are children under the age of 13 active in Habbo?[15] We do not take artistic quality criteria into account and concentrate on daily creative activities distinguished by the Habbos themselves.

Research into everyday creativity can involve various aspects, such as creative personalities or characteristics of these people, creative products or creative processes, or the skills that children require to be creative. In this exploratory research we approach creativity as *activity*, creativity as in the creation of something.[16] Children in Habbo can be creative because they *do, organise* or *make* certain things that are, to a certain extent, new, surprising or unexpected. This approach allows Habbos to be creative to a greater or lesser extent in their virtual world. Combining various existing elements into something new is, for example, less creative then setting up, from scratch, something that has never before existed in Habbo.

CREATIVE ACTIVITIES

The creativity scale in virtual worlds has three activities or roles that users can assume.[17] In increasing order of creativity these are: adapting, initiating/organising and creating. The higher the activity is on the scale, the greater the capacity of children to act independently in their environment and the greater the creativity of the users.

Adapting – To the left on the scale can be found 'adapting'. This involves a limited degree of creativity in which children can make their own combinations using existing elements. For example, putting to-

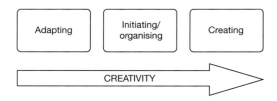

Figure 4.3 • Creativity scale in virtual worlds

gether an avatar from existing elements and designing a room using existing attributes.

Initiating / organising – The second step on the scale of creativity is 'initiating/organising'. Besides participating in existing activities in the virtual worlds, children can also initiate activities themselves. They can implement their own ideas. They make use of existing elements from the virtual surroundings to do this, but present them in a new combination or with a new meaning. Solving certain obstacles stumbled upon in their virtual world is also an example.

Creating – In this (most creative) activity, something new is built up from nothing. Children can, of course, have been inspired by certain elements (from the virtual world) but, as such, what they created did not yet exist.

This creativity scale can be used on the things children do in Habbo.

4.4 • CREATIVITY IN HABBO

In comparison to the virtual worlds of adults and compared to open worlds as Whyville and Shidonni, children in Habbo do not have any possibilities of programming or creating objects from scratch. It is actually forbidden to add new elements to the virtual world. Their creative freedom, on the face of it, looks restricted.

Yet creativity within Habbo most certainly plays a role. According to the manager of Habbo in the Netherlands, in the age range of most Habbos 'expression' and ' looking for an identity' (and a role in society) are important. Creativity is a means for self-expression, a way to stand out and achieve respect.[18]

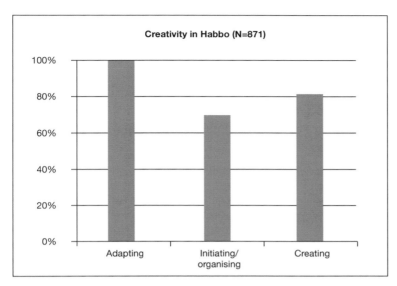

Figure 4.4 • Percentage of Habbos that are creative (N=871)

In 2009, a survey was conducted in the Netherlands to determine, among other, what the creative possibilities of Habbo are. It appeared that almost 70% of the children in Habbo consider Habbos to be creative.[19] In addition, 58% of the Habbos who participated in the survey said that they were surrounded by many creative Habbos. However, as many as 35% think that the freedom in Habbo is restricted and that they cannot do everything they would want to. The activities of young Habbos are analysed below, taking into account the above mentioned creativity scale. They then appear to be creative at different levels (see Figure 4.4).

ADAPTING

"Well, to make mine hotel me needed much furni and kreativiti" (quote Habbo)

Children can be creative in Habbo by designing their own avatar and furnishing their own room. They can choose from various heads, bodies, skin colours, hairdos and clothing styles. They can also furnish their own room. Two-thirds of the Habbos say they often change the design of their room. Technically it should be possible for Habbos to make their own clothes and furniture. The Habbo organisation has, however, chosen not to allow this. In this way the hotel remains manageable.

"I made a super special disco for someone and she was so touched, she liked it so much :D" (quote Habbo)

Are Habbos satisfied with the way in which they can be creative at this? Almost 65% of the Habbos younger than 13 years of age who took part in the study felt that they had insufficient options to change Habbo. And as many as 93% of the surveyed Habbos said that they wanted to design their own Habbo. So, they are nowhere near satisfied with the options offered.

"Trying to copy the White House" (quote Habbo)

The fact that there are limited options available to create their own avatars does not mean that children cannot come up with creative ways to manage their Habbos. Almost one-third, for example, say they have more than five avatars. In online discussion groups in 2007 it was asked why they do this. Some Habbos use their different avatars to collect free credits (for example, at Christmas). Other Habbos have avatars as reserves in the event they get hacked. If that is the case they can change to another character. Some Habbos said that they sometimes use a different account to go undercover. They do not feel like being 'themselves' for a while. They explain it as follows: "you have a certain image as Habbo :P. so if you go around as another Habbo you lose the other image for a while."

"Made an elephant with small tables" (quote Habbo)

This also applies to going about with 'furni'. In order to buy furniture Habbo Credits are necessary. Most visitors, however, do not spend any money in the hotel. They do not have any or their parents do not allow them to spend it. Still, most Habbos do have furniture; 30% claims they even have more than 100 furni. They gather this furniture in all sorts of ways (see Figure 4.5). Sometimes they get furni as a present, from Habbo, or from a friend. They sometimes have to do something in return, like promoting a specific room. Bartering is also actively takes place between the Habbos.

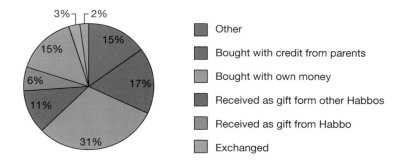

origin most meubi (N=871)

- Other
- Bought with credit from parents
- Bought with own money
- Received as gift form other Habbos
- Received as gift from Habbo
- Exchanged

Figure 4.5 • Origin of furniture in Habbo (N = 871)

"An entire show, where everyone could join in with Habbo. But there was just 1 rule… You had to improvise everything ;P "(quote Habbo)

Another way of being creative in Habbo is by organising activities and events. Activities invented by Habbos are often small-scale. For example, fashion shows and hunts. The activities are sometimes based on things that happen in real life, such as Queens Day, Christmas, Hallow-

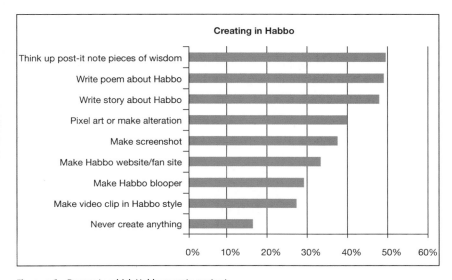

Figure 4.6 • Degree to which Habbos create content

een or commemoration of the dead. Almost 70% of Habbos have organised an activity in Habbo. In addition, 90% of Habbos aged less than 13, claim to have taken part in an activity organised by a fellow Habbo.

"My army trying to keep habbio [sic] safe" (quote Habbo)

Another interesting phenomenon is the way in which Habbos think up their own jobs. Most of the jobs carried out by Habbos are thought up by other Habbos and reflect the jobs that people have in real life. Popular jobs in Habbo are: models, advertising (for example, for a certain room), police, help desk staff (Habbos voluntarily help new Habbos), babysitter and even a Maffia member. More than 62% of the Habbos aged less than 13 say that have had a job at some stage.

CREATING

"A poem of mine has in the past been published in the newsletter – I thought it was original J :) and creative" (quote Habbo)

'Creating' in Habbo means creating content. As many as 84% does this at least once a year. Figure 4.6 shows the type of activities concerned.

Habbos think up post-it note pieces of wisdom, and write poems and stories to do with Habbo. They also create Habbo pixel art or alterations (adaptations to images in Habbo pixel style). For example, they take a picture of their favourite pop star and use it to make a Habbo pixel character. Other types of content creation are the making of screenshots, thinking up Habbo bloopers and making video clips in Habbo style. Outside of Habbo users also make their own fan sites. They collect information and news related to Habbo.

"On Habbo itself nothing but I make pixel art" (Habbo)

A mere 16% of Habbos never takes part in these types of creativity. It is striking that Habbo itself, apart from the personal pages that Habbos have, offers nowhere to document these expressions of creativeness. The alterations cannot, for example, be introduced into the virtual world itself because of the limited freedom within the hotel. Habbos therefore often send their creations to the Habbo management who sometimes publish them in the Habbo newsletter. If competitions are organised, to make films, for example, a place is created to show the entries, say on the Habbo Netherlands homepage.

In addition to adapting, initiating/organising and creating, there is another special form of creativity used in a creative way by children to get round the strict rules within Habbo. As mentioned earlier, Habbo offers little space to express creativity. Some Habbos disagree with this and try to get round the rules. They nevertheless program items that are not allowed to exist in the safe and controlled Habbo environment. Or they think up tricks to by-pass the word filter so as to be able to swear. Or they pilfer other people's password (scamming), or hack their accounts. Habbo is continuously waging a battle against these creative spoilsports. Most Habbos refer to scammers (people using excuses to get to other people's furniture or passwords) and hackers (people using technical means or software to break into your computer to steal your Habbo password) as the greatest nuisance.

4.5 • CONCLUSION

Critics believe that virtual and commercial virtual worlds do not allow for creativity. In line with this you could say that children in virtually existing worlds can only be apparently creative. The analysis of everyday creativity in Habbo, given above, shows that this is untrue.

At the lowest level of creativity all Habbos are creative in adapting their character and designing their room. As the options are limited, Habbos invent alternatives. They often have several avatars and have creative ways of obtaining furniture. Even pre-programmed choices can be dealt with in an entirely personal way.

A large group of Habbos is active on the second level of creativity – initiating and organising their own activities. For example, they think up competitions, fashion shows, plays and games. Because Habbo has many users, the value of thinking up own activities is great. There are always Habbos in the neighbourhood who want to enter a self-made competition, especially if there are prizes to be won. This also applies to the inventing of jobs. Habbos observed that they have many creative fellow Habbos around them.

At the highest level of creativity – the creating of new things – many activities also take place in contrast to what is often thought. The reason for these misunderstandings is that they often concern activities that are not visible on Habbo itself, such as the making of pixel art (and publishing it elsewhere), writing poems or stories about Habbo and the making of fan sites. These activities actually form a link between the virtual Habbo world and the world outside (physical or online).

Habbo offers children all kinds of ways to be really creative in an everyday manner. Habbos partly follow pre-defined pathways. But they also appear to be creative in ways not anticipated by the makers. If the options are limited within Habbo, the children think of their own solutions, for example by being creative with the Habbo theme offline. Or they get round limiting Habbo rules in a creative way. In this manner the creativity in Habbo is not 'just apparently existent'.

The assumption that children who come to Habbo are mainly raised to become consumers and given no possibilities to be creative, is a considerable underestimation of reality.

NOTES

1 The quotes in this chapter are illustrative. They have been copied from the literal reports of the discussion groups that took place in Habbo and the results of the survey in which the Habbos participated in 2009. In this English edition, the quotes have been translated.

2 Malaby, T. (2005). Parlaying Value: capital in and beyond virtual worlds. University of Wisconsin-Milwaukee. Downloaded on 17 October 2009 from SSRN: http://ssrn.com/abstract=871851.

3 That is why these environments are also called MUVEs : Multi User Virtual Environments. Koenraad, T. (2007) 3D virtual worlds for the MVTO: focus on virtual 'language village'. Downloaded on 17 October 2009 from http://www.koenraad.info/publications/publications-in-dutch.

4 In 2009 the number of virtual worlds for children aged between 8 and 12 was estimated to be 88, by Van Bruggen. For children less than 7 years of age there are 72 virtual worlds available according to Van Bruggen. See: Van Bruggen, W. (2009). *Virtuele werelden voor de PO doelgroep.* Downloaded on 16 October 2009 from http://virtueleomgevingen.nl/muvepo.

5 Virtual Worlds News (2008). Dubit: 70% of Tweens Respond Positively to Branding in Virtual Worlds. http://www.virtualworldsnews.com/2008/10/dubit-70-of-twe.html (consulted on 17 October 2009).

6 Habbo in the Netherlands is a joint venture between Sulake and the Telegraaf Media Group. The Dutch Habbo opened up in February 2004 and within 2,5 years, nearly 4,5 million accounts created. In 2009 this number stands at 137 million worldwide, of which 13,4 million in the Netherlands. Around 950,000 children visit the Dutch Habbo every month. Most of them visit Habbo twice a day, and the average total daily visiting time is 40 minutes. Most Habbos are, according to Sulake, roughly 14 years old, but they usually start playing on Habbo much earlier. Of the Habbos that participated in the 2009 research, more than 78 percent of Habbos under 13 had been coming to Habbo for more than a year, nearly half of them more than two years.

7 Some of the virtual worlds discussed here are not easy to visit for Dutch children. This has to do with the language. Whyville and Club Penguin, for example, are only in English. Also Barbiegirls and Pixie Hollow are English language websites. Some services, such as Taatu, Habbo and goSupermodel, offer their virtual world in several localised editions, among which a Dutch version. This makes it easier for Dutch children to visit these worlds.

8 For more information on Whyville see Gallas, C. & Sun, J. (2007). Whyville. Downloaded on 16 October 2009 from http://www.cathleengalas.com/projects/galas_sun_whyvillechapter.pdf and Knapp, L. (2007). Why is Whyville a hit? It's safe. The Seattle Times, 17 February 2007. Downloaded on 16 October 2009 from http://seattletimes.nwsource.com/html/businesstechnology/2003576142_ptgett17.html.

9 Depending on the way you purchase your credits and how many you buy.

10 See Livingstone, S. & Bober, M. (2004). *UK children go online: surveying the experiences of young people and their parents [online]* Londen: LSE Research Online. Downloaded on 29 September 2009 from http://eprints.lse.ac.uk/archive/00000395;
McGuire, M. (2005). Ordered communities, *M/C Journal*, jrg. 7, (nr.) 6.

11   Definition of creativity as given in Sternberg, R.J. and Lubart, T.I. (1999).
     Handbook of Creativity Cambridge: Cambridge University Press. Among creativity
     researchers, this is the most often used definition.

12   Kaufman, J.C. & Baer, J. (2004). Hawking's haiku, Madonna's math: why it is hard
     to be creative in every room of the house. Sternberg, R.J., Grigorenho, E.L.,
     Singer, J.L. (eds.) *Creativity: from potential ro realization*. Washington: American
     Psychological Association.

13   See Csikszentmihalyi, M. (1996). *Creativity: flow and the psychology of discovery
     and invention*. New York HarperCollins.

14   Van den Broek, A., De Haan, J. & Huysmans, F. (2009). *Cultuurbewonderaars en
     cultuurbeoefenaars; trends in cultuurparticipatie en mediagebruik* (Culture admir-
     ers and culture practitioners; trends in culture participation and media use). The
     Haque: SCP.

15   This chapter is based on a case study of a PhD thesis on changing roles of users of
     online media entertainment services in 2007-2007 as well as an exploratory re-
     search in children's creativity in Habbo.nl. It was carried out in 2009 and 871
     Dutch children aged under 13 participated. This group consisted for 45 percent of
     boys and 55 percent girls. It should be noted that the Habbos that took part in the
     study, can largely been considered as belonging to the active group of Habbos.
     They are generally online for a longer time than the average Habbo, are Habbo
     Club member more often and spend more money on it. This research is a follow
     up on a PhD thesis into changing roles of users, which studied the Dutch Habbo
     Hotel case in 2006.

16   This approach is lightly based on the one by Becker in the art sociology, that sees
     art as a total of activities (Becker, H.S. (1982). *Arts Worlds*. Berkeley, Los Angeles,
     London: University of California Press) and research by the author into changing
     roles of users in online media entertainment services (see also next foot note).

17   For more literature on the roles of users, see earlier research by the author, for
     example Slot, M. (2008). Expanding user roles in digital television In: Urban, A.,
     Sapio, B. & Turk, T. Digital Television Revisited. Linking users, markets and
     policies. Workshop proceedings COST Action 298 'Participation in the Broadband
     Society', pp.108-122 and Slot, M. and Frissen, V.A.J. (2007) 'Users in the 'golden'
     age of the information society' *Observatorio (OBS*) Journal*, (nr.) 3, pp. 201-224.

18   Personal communication with Mark Stockx, country manager Habbo NL, Septem-
     ber 2009.

19   This is brought forward by the study among Habbos in 2009, but also by previous
     research. See: Slot, M. (2009). Exploring user-producer interaction in online com-
     munities; the case of Habbo Hotel. *Journal of Web Based Communities,* year. 5,
     (nr.) 1, pp. 33-48.

"I dare to be honest with my internet friends"

Isa (12) and Leonore (12) met last summer at the campsite. As it turned out, they were both fans of goSupermodel and added each other. They are more frank with their online friends.

Isa: "I have really good friends on goSupermodel. I can usually be a lot more honest with them than with friends I know in real life. I tell them everything. About how I feel. I discuss this in more detail on the internet. I also dress differently there. My hair is wild, with lots of colours everywhere. Last year I also wore those kinds of clothes at school but I stopped doing that because people gossiped and made blunt comments. On goSupermodel it doesn't matter what you wear as long as you are nice."

"I write on the forum about things that interest me, such as friendships or boys. It is because you are anonymous, you can be yourself. Online I am often hyperactive, happy and funny. In real life I'm sometimes not like that at all. I'm different from many girls in my class. I'm not too interested in brands but prefer to be around people I love. That is why I sometimes don't say what I really feel because if they don't agree it makes me feel bad. People on the forum also go on at you sometimes but in real life it hurts more.

I'm actually slightly addicted to the computer. I don't know what I'd do in my spare time if I didn't have the internet. I think I'd then tell even less about myself."

Leonore: "I've met some girls on goSupermodel in real life, like Isa. But most of them I only know from the internet. In real life you are often prejudiced about how someone looks. Its much nicer to talk to someone with whom you don't have this. I talk more easily, not because I've never really seen them, but more because they can't see my facial expression at the time."

"I can always get support from my friends on goSupermodel. They are very nice to me and we almost never have an argument. Only occasionally, if you say something they don't like. I also behave differently on goSupermodel. Sometimes I don't dare say something to friends because they think I'm irritating. They never say that to me on goSupermodel. And if I don't understand something, my friends sigh because it happens quite a lot. But on goSupermodel they just explain it to me. The nice thing is that I can also take my time in thinking what I want to say. When I talk pause a lot. This is not a problem online."

"My doll on goSupermodel doesn't really look like me. I don't dress at all like her. She wears long glittery dresses. I'd never wear them in real life. Trousers, a shirt and a nice necklace, that's me."

Gonnie, Isa's mother: "She has lots of fun with online friends. I always hear her laughing a lot. But I still think it's some kind of poverty having to communicate via the net. You have to wonder why girls nowadays always have to criticise each other. Why do they set such high standards for each other?"

"We have just reduced the computer time to two hours. The computer turns off automatically after 2 hours. Making other deals never worked out and always ended in arguments. If she spends hours at the computer after dinner and having done her homework she wouldn't be able to go to bed until 10:30 pm. That's far too late. My experience is that you can't expect children to be able to go about this sensibly by themselves."

5

Communicating online

Patti Valkenburg
Jochen Peter

MSN and social network sites, such as Hyves, appear to have been made for young teenagers. Why is this the case? In this chapter we try to find an answer to the question why young teenagers, 10- to 13-year-olds, are so attracted to these types of online communication. We will also make an estimate of the opportunities and risks of online communication technology for this age group, based on research we have carried out over the past years. Finally, we will make some policy recommendations in order to maximise the opportunities of online communication for this age group, while minimising the risks.

This chapter focusses on young teenagers and is based on information gathered among 10- to 13-year-olds. This does not mean that 6- to 9-year-olds do not use the internet for online communication nor that the internet has no opportunities and risks for this group. However, too little research has been carried out on online communication among younger children to be able to make founded judgements on this.

5.1 • NEED FOR COMMUNICATION WITH PEERS

The early teenage years are a period of dramatic physical, cognitive and emotional change. For almost all boys and girls this is the start of puberty, for girls usually one or two years earlier than for boys. Strictly

Figure 5.1 • Social network sites, such as Hyves, offer teenagers a relatively easy way to be constantly in touch with friends and peers.

speaking, puberty refers to a period in which teenagers start to be able to sexually reproduce themselves. But more generally, it refers to the physical changes that take place between the period of childhood and adolescence, including the rapid growth in height and the increase in sex hormones. This physical and sexual maturation is accompanied by important cognitive, social and emotional changes. Together, these changes have a fundamental effect on how teenagers see themselves and others.[1]

One of the main developmental tasks in the teenage period is gaining autonomy from parents and developing a new, mature identity. To get this mature identity, many new roles have to be learnt and practised. This mainly takes place by interacting with peers. There are two ways in which teenagers practise new roles. In the first place by binding themselves to a peer group, a group of peers who share a number of important interests, and secondly, by forming close friendships.

To become part of a peer group and to form close friendships, teenagers need to acquire at least two skills: they have to learn how to present themselves in the proper way (self-presentation) and they have to learn how to reveal intimate personal details (self-disclosure). Self-presentation and self-disclosure are related concepts but they do differ from each other:

- Self-presentation indicates the way in which individuals present their identity to others;
- Self-disclosure is the degree to which they communicate intimate details about themselves to others.

Self-presentation and self-disclosure are both very important for the cognitive, emotional and social development of young teenagers. They can, for example, use self-presentation and self-disclosure to check how their own knowledge and views compare to those of peers. They can also get emotional support from friends by showing their vulnerable side (by self-disclosure). Finally, appropriate self-presentation and self-disclosure have a positive effect on the formation and quality of friendships.[2]

Self-presentation and self-disclosure are not automatically developed; they are skills that have to be practised. Friendships can, for example, only develop properly if both friends exchange well dosed intimate details. Self-disclosure often stimulates reciprocal self-disclosure, and these reciprocal revelations ensure that friends get to know and appreciate each other better. It can be counter-productive if someone does not reveal enough about himself or herself. This can be off-putting. If teenagers are not capable of adequately learning self-presentation and self-disclosure it can have serious emotional and social consequences, such as social exclusion, loneliness and depression.

5.2 • THE APPEAL OF ONLINE COMMUNICATION

It is, therefore, not surprising that young teenagers have an intense need for self-presentation and self-disclosure and, as a result, want to constantly be in touch with peers. After all, practising self-presentation and self-disclosure is functional. The need for self-presentation and self-disclosure is, in itself, reason enough to visit the internet for communication purposes. However, there are two other factors that clarify the great need of young teenagers to communicate online.

The first factor is of a practical nature. Particularly younger teenagers have limited possibilities for face to face communication with peers. They are, after all, not allowed to go out after school (or in the evenings) as often as older teens. Particularly for younger teens the internet therefore offers a relatively easy environment to be in constant touch with friends and peers.

A second factor is that young teenagers can be shy and unsure of themselves in their self-presentation, not least because of the great physical changes they are going through. They want to be in touch with peers but feel hampered by feelings of insecurity. They are, therefore, especially interested in an environment in which they can relatively safely practise their self-presentation and revelation.

The internet offers such an environment. First of all, online communication is characterised by less auditive and visual information. Even though many teenagers use a webcam, less is visible during online communication than during face to face contact (for example, pimples, blushing, etc). Many young teens appreciate these concealing properties of the internet. Our research shows that young teenagers who value the fact that others cannot see them on the internet, feel more freedom and uninhibited to reveal intimate details about themselves, for example, about their concerns, crushes and things they are ashamed of.[3]

Another reason that self-presentation and self-disclosure on the internet are more interesting for young teens is the fact that posing direct questions is more common. Different studies have shown that teenagers ask each other more direct questions on the internet than during face to face contact.[4] This is probably considered to be less rude (as you often do not see each other). The possibility of being able to ask direct questions, and more or less getting away with it, has important consequences for the reciprocity of online communication. If direct questions are not considered to be rude, a communication partner is more likely to give an answer. It is, however, a well-known socio-psychological process that self-disclosure leads to reciprocal self-disclosure. And this is probably the reason that teenagers not only exchange more, but also more intimate, details on the internet than during physical encounters.

5.3 • OPPORTUNITIES

Online communication offers young teenagers a number of important opportunities. Over the past years we have published a number of studies that show that the internet offers teenagers at least four opportunities socially. Online communication has a stimulating effect on:
• Formation of friendships [5];
• The quality of existing friendships [6];
• Self-confidence [7];
• Social competence[8].

These four social opportunities can be explained by the possibilities that the internet offers teenagers to practise and improve their self-presentation and self-disclosure.

THE EFFECTS OF ONLINE SELF-DISCLOSURE

In various studies we have shown that using MSN can lead to more intimate self-disclosure.[9] This has far-reaching consequences. First of all, it appears that teenagers who reveal more about themselves on the internet, find it easier to make friends online.[10]

In addition, online self-disclosure also has a positive effect on the quality of existing friends. One of our most recent studies, in which we followed approximately 800 teenagers aged between 10 and 17, showed that 10- to 12-year-olds who reveal a lot about themselves via MSN, had higher quality friendships after six months than teens who did not.[11] Socio-psychological research has, over the past decades, shown several times that self-disclosure is a very important determinant of the quality of friendships. This particularly applies to young teenagers.

Young teenagers who find it easy to express themselves intimately receive more advice and social support from their environment. They have friendships of higher quality and are often also socially more competent and happier. These results have, to date, been found for offline self-disclosure. It is important to know whether the positive effects of self-disclosure also apply to online self-disclosure.

THE EFFECTS OF ONLINE SELF-PRESENTATION

The possibility for teenagers to practise their self-presentation via the internet, for example on social network sites, also offers opportunities. Our research shows that practising self-presentation on social networks not only increases the self-confidence of young teenagers but also the social competence.[12]

Approximately 20 to 50% of teenagers regularly experiment with their identity on the internet. They pretend to be older, prettier, nicer, and flirtier than they really are. They have at least three motives for doing this:
• To try out aspects of their identity;
• To make friends more easily;
• To overcome their shyness.

Teenagers, both young and older, who experiment with their identity more often, have more self-confidence and are more socially competent. They make friends and stand up for themselves more easily. This effect of identity experiments on the internet results from the fact that teenagers who do this often communicate with a greater variety of communication partners when they are online. For example, they talk to more people with social, cultural or intellectual backgrounds other than their own.[13] This probably gives them more opportunities to train their social competences.

Figure 5.2 illustrates the four social opportunities. The first two (development of friendships and the quality of existing friendships) are caused by internet's potential to stimulate self-disclosure among teenagers. The last two (self-confidence and social competence) are caused by the greater possibilities offered by the internet for self-presentation.

5.4 • RISKS

It goes without saying that the internet is not the heavenly paradise where all teenagers are helped. In the first place, it is important to note that the opportunities that the internet offers do not apply to all online communication technologies or to all teenagers to the same extent.[14] The positive effects of online communication on the quality of existing friendships, for example, only emerge if teenagers use the internet mainly to communicate with existing friends. MSN is an example of a technology used by the most teenagers to communicate with existing friends. The positive effect of MSN does not, however, occur if young teenagers mainly use MSN to communicate with strangers.[15] The teenagers who do this form a small, but conceivable minority. The positive effect of online communication on self-confidence also only applies if teenagers have created their online profile, for example on social network sites, in such a way that they receive mainly positive feedback on

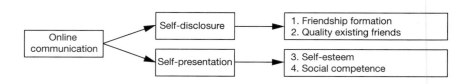

Figure 5.2 • Social opportunities

it. The self-confidence of teenagers who do not manage to do this is not stimulated.

In addition, online communication also has real risks. As is the case with opportunities, risks often stem from the disinhibiting function of the internet. The more open behaviour on the internet, also called disinhibition in the literature, can result in positive effects (see above) but it can also have negative consequences. On social network sites, such as Hyves or Facebook, for example, teenagers may expose too many details about themselves. More than two-thirds of the 12- to 13-year-olds are currently connected to one or more social network sites. Many young teens give a lot of information about themselves on these sites. A quarter of the 12- to 13-year-olds mention their surname to everyone. Also, 25% of the children in this age category mention the school they are at. And 5% publicly mentions their sexual preference. In addition, 2% in this age category had at some stage in the past six months spread a photo of themselves via the internet on which they were partly or totally nude. Our latest information shows that an estimated 2% of the young social network users are fully transparent in that they expose everything about themselves. Initially, this percentage does not seem alarming but if you realize that 2 million teenagers surf around in the Netherlands, it could mean that as many as 40,000 teenagers could run into problems because they have exposed too much of themselves.

Finally, young teenagers can get into problems because of the uninhibited behaviour of other internet users. They can easily abuse the reduced auditive and visual information of online communication. Young teenagers are regularly confronted with this. Some figures from our latest survey (autumn 2008):
• 18% of 12- to 13-year-olds said they had been bullied on the internet in the past six months;
• 13% had been harassed online in the past six months;
• 6% had met someone on the internet who wanted to talk about sex with them;
• 8% had experienced that others had spread intimate information on them.

Young teenagers are the most vulnerable group with regard to online communication. This is due to a combination of factors.

First of all, having many contacts is particularly important in the teenage years as it gives status in the peer group. In the later teenage years, the quality of friendships becomes more important than the number of friendships. This great need for contacts means that teenagers are more likely to seek contact with strangers on the internet.[16]

Secondly, young teenagers can suffer from being irritable, impulsive and aggressive as the result of hormone fluctuations during puberty.[17] This can lead to provocative behaviour on the internet which, in turn, can provoke negative reactions from others. It is, for example, striking that young teens get negative feedback on their online profile more often then older teens. They provide information on these profiles whose effects they cannot assess properly.

Thirdly, young teenagers also experiment with their identity (on the internet) more often than older teenagers. They often pose as someone older and prettier than they really are. They particularly do this to get in touch with strangers.[18]

In summary, especially young teenagers seek the contact of strangers on the internet and they in particular experiment with their identity on the internet. This happens exactly at the time when they are most vulnerable. Teenagers explore their boundaries and experiment with them just because they do not know their boundaries very well yet. Their identity is still relatively unstable and negative experiences can be particularly painful.

POLICY

What should the policy be with regard to teenagers and online communication? We have shown that, particularly for teenagers, the internet has risks. Never before have media and technologies had such a direct effect on the vulnerable identity of young teenagers. This creates new questions about the education of teenagers.

What can we do about this? No one will deny that the upbringing of children is primarily the responsibility of the parents. But it is now more difficult for parents than it used to be to fulfil their gatekeeper function.

Although research has shown that parents can usually adequately counter the negative effects of the media, most of these strategies are only effective for young children but far less so for teenagers.

Research suggests that parents have less knowledge of the internet than their children.[19] They therefore have an extra need for information on internet-specific education and the effect of the internet on their children. The concern of parents is often concentrated on their lack of technical knowledge. However, they do not or not sufficiently realise that it usually concerns problems that also existed before the internet era, such as: crushes, insecurity, social exclusion, sexual solicitation, bullying and aggressive behaviour.

Our most recent study showed that many of these problems, including bullying, still occur less often on the internet than in the real physical life. Good advice for parents on these subjects can usually be found in the literature on education in which the social problems in the adolescent period are discussed.

PARENTS, EDUCATION, INDUSTRY AND GOVERNMENT

Parents are primarily responsible for protecting their children and making them media-literate. But the developments are rapid. Parents are not always media-literate themselves and therefore need to be informed. Parents should be able to hand over part of their responsibilities to others. There are three parties that qualify:

• Education should take responsibility for making children media-literate. Media literacy programmes should be focused on the social aspects of the internet as education in this area is most important for this age group.

• The industry (internet providers, media providers, etc.) have the responsibility to provide transparent information and to keep to the regulations, laid down by the government or a trade organisation. The College Bescherming Persoonsgegevens (The Dutch Data Protection Authority), for example, published a guideline in October 2007 with special privacy guidelines for youngsters under the age of 16. Internet providers have to comply with these guidelines.

• The government has to see to it that all other parties (parents, education and technology providers) can optimally realize their specific responsibilities. The past has taught us that discussions on and the regulation of social and ethical aspects of technological developments often lag behind the technological developments.

The government's policy with regard to regulations has long been characterised by restraint. This pedagogic shyness applies to both the protection of children and to making children media-literate through media information in educational programmes.

Over the past years, it has become apparent that joint regulation, with partial involvement by the government, works better than complete self-regulation. This can be clearly seen in the development and further expansion of the Dutch media classification system *Kijkwijzer*. Money and ethics do not really go together and enforcing the *Kijkwijzer* without the government would be much more difficult.[20]

A society in which the development of an identity has become a deliberate choice calls for investment by the government.[21] These investments should certainly apply to the protection of young teenagers on the internet and to making them media-literate.

NOTES

1   Steinberg, L. (2008). *Adolescence.* Boston: McGraw Hill.
2   Valkenburg, P.M., & Peter, J. (2009). The effects of Instant Messaging on the quality of adolescents' existing friendships: A longitudinal study. *Journal of Communication, 59,* 79-97.
3   Schouten, A.P., Valkenburg, P.M., & Peter, J. (2007). Precursors and Underlying Processes of Adolescents' Online Self-Disclosure: Developing and Testing an "Internet-Attribute-Perception" Model. *Media Psychology, 10,* 292-314.
4   Antheunis, M., Valkenburg, P.M., & Peter, J. (2007). Computer-mediated communication and interpersonal attraction: An experimental test of three explanatory hypotheses. *CyberPsychology and Behavior, 10,* 831-836.
5   Peter, J., Valkenburg, P.M., & Schouten, A.P. (2005). Developing a model of adolescents' friendship formation on the Internet. *CyberPsychology & Behavior, 8,* 423-430.
6   Valkenburg & Peter (2009). See note 2.
7   Valkenburg, P.M., Peter, J., & Schouten, A.P. (2006). Friend networking sites and their relationship to adolescents' self-esteem and well-being. *CyberPsychology & Behavior, 9,* 584-590.
8   Valkenburg, P.M., & Peter, J. (2008). Adolescents' identity experiments on the Internet: Consequences for social competence and self-concept unity. *Communication Research, 35,* 208-231. [IF 1.48].
9   Bijv. Antheunis et al (2007). See note 4; Valkenburg & Peter (2009). See note 5.
10 Peter et al (2005). See note 5.
11 Valkenburg & Peter (2009). See note 2.
12 Valkenburg & Peter (2008). See note 8.
13 Valkenburg & Peter (2008). See note 8.
14 Van den Eijnden, R. et al, (2008). Online communication, compulsive Internet use, and psychosocial well-being among adolescents: A longitudinal study. *Developmental Psychology,* 44, 655-665.
15 Valkenburg, P.M., & Peter, J. (2009). Social consequences of the Internet for adolescents: A decade of Research. *Current Directions in Psychological Science, 18,* 1-5.
16 Peter, J., Valkenburg, P.M., & Schouten, A.P. (2006). Characteristics and motives of adolescents talking with strangers on the Internet. *CyberPsychology & Behavior, 9,* 526-530.
17 Steinberg, L. (2008). See note 1.
18 Valkenburg, P.M., Schouten, A.P., & Peter, J. (2005). Adolescents' Internet-based identity experiments: An exploratory survey. *New Media and Society, 7,* 383-402.
19 Duimel, M., & De Haan, J. (2007). *Nieuwe Links in het gezin: de digitale leefwereld van tieners en de rol van hun ouders* (New links within the family: the digital world of teenagers and the role of their parents. The Hague: The Netherlands Institute for Social Research (SCP).
20 This observation has come about as a result of the first author's membership of the scientific committee of NICAM (Netherlands Institute for the Classification of Audio-visual Media), in which she has been jointly responsible, since 2001, for the development of the Kijkwijzer.
21 Valkenburg, P.M. (2005). *Schadelijke media en weerbare jeugd: een beleidsvisie 2005-2010.* (Harmful media and tough youngsters: a vision statement). Amsterdam, Amsterdam University. Report commissioned by the Committee Media, Youth, and Violence.

"I know all my Hyves friends in real life too"

Sarah (7) and Yin (8) are best friends and classmates. They both have a Hyves profile.

Sarah: "All my friends were already on Hyves and I also wanted to join. But my mother didn't allow it at first. When I asked my dad, he said yes. Ha ha, that was clever."

"I know all my sixteen friends on Hyves also in real life, from school, etc. I send them scribbles. And I also use Buddypokes a lot. These are dolls that look like you. Then you can ride together on a scooter, for example, or Yin reads a story and I fall asleep. Then she's angry, that's very funny."

Yin: "I also like Buddypokes. You can cuddle each other and ask: did you like cuddling? But a Buddypoke also kicked me once."

Sarah: "Yes, someone made a judo Buddypoke and it kicked. I then removed it."

Yin: "I think I have about 17 friends. Both boys and girls. My father only allows me to have friends that I know. That's fine, otherwise people I don't know can see my photos and I don't want that."

"I have two photos on it now. I'd like to put more photos on it but am not sure how to do this. So I ask my mother. She also made my profile."

Sarah: "Mummy also helps me a bit but I did most of it myself. I have now made a very cute little hamster that devours a big bad wolf. I thought this up myself, it was so funny. I also have films, of myself and of High School Musical."

Yin: "I often send scribbles. My friend had a sore knee and I scribbled: how are you? I also often make my Hyves page pretty. I think I must spend at least an hour on it everyday."

Corien, Yin's mother: "I initially didn't want her to go on Hyves. There are so many things that I find more important for her than the number of friends she has. I first wanted her to become aware that friends of friends can also see what you're doing on Hyves. That you shouldn't put certain things on there, things you would normally not share with others either."

"I'm also on Hyves, she likes this. And since the holidays she now has her own page. She added me and yesterday I received a Buddypoke and a big cuddle from her that said: 'I love you, mummy'. So sweet. Those Buddypokes are very nice, they can cuddle each other and pimp themselves with clothes and hair and eyes and show their feelings to others."

"She scribbles short notes, like: 'how's things?' And she uses many smileys. I haven't seen her pour out her heart yet on Hyves. She knows that if she doesn't agree with a friend she should tell her to her face and not via the computer."

"If it were up to her she'd go straight to the computer when she comes home from school. She now has Hello Kitty glitters on her page. She's more interested in the gadgets than the actual communication."

6

Hyves

Marion Duimel

Hyves is by far the largest social network site (SNS) in the Netherlands. SNS is a collective term for all kinds of websites focusing on social relations. More precisely: online environments in which people can make a profile of themselves with all sorts of information and make links with others by including them in their contact list. Everyone can glance at the profiles through their own list and that of others.[1] Profile pages can be public or only visible to acquaintances. Examples of other network sites are: Facebook, MySpace and LinkedIn. The latter websites attract a mainly older audience and are less fascinating for children. For Dutch children Hyves is the absolute favourite.

For those who would like to know more about Hyves, particularly about its use for children, they could ask themselves the following questions: how many children are actually on Hyves? Do they really have 'hundreds of Hyves friends', as is regularly stated? And if so, who are these friends? How important is it to have as many friends as possible? What do 8- to 12-year-olds consider the most important features of Hyves? Is it mainly about 'scribbling' (leaving messages for each other) or do they prefer to play games?

Hyves offers many communication possibilities. These opportunities are also discussed in Chapter 5 (Online Communication). But what are the risks? Privacy is a topic widely discussed. If children do not protect their personal information, photos and scribbles they can be seen by everyone. Do parents guide their children on safe use of the website and how many parents set rules for its use? Besides, children sometimes dare to say more on Hyves than when face to face. This can result in undesired scribbles or even arguments. To what extent do children experience this on Hyves? Which children are most vulnerable?

All these questions will be answered below on the basis of research carried out in July 2009 among 524 children aged between 8 and 12, by market research company Qrius[2] and commissioned by the Stichting Mijn Kind Online (My Child Online Foundation).

6.1 • HYVES IN BRIEF

Hyves (derived from *beehive)* was established in 2004 by three friends. Inspired by social network sites in America they decided to adapt the concept to their own wishes and ideas and introduced it in the Netherlands.[3] After a short running-in period, students picked it up and from 2005 onwards the number of members grew explosively. In March of that year there were already 100,000 members and by August a million profiles were counted. In December 2007 the site welcomed its 5 millionth member.

The popularity of Dutch network sites Cu2 and Sugababes/Superdudes, initially the profile sites for young Dutch teenagers, decreased visibly in the period 2006-2008 and an enormous number of members moved to Hyves. In 2010 Hyves already has more than 9 million profiles. This is, of course, not the same as 9 million active members. There are, for example, profiles which were once made but are no longer used, double profiles and faker's profiles, in other words sham profiles made by others. How many real members there are is unclear. But according to Hyves co-founder Raymond Spanjar approximately two-thirds log in once a month. That would mean 5.5 million members.[4]

6.2 • AGE

With this many members it is not surprising that most generations are represented on Hyves. When searching by age you can even choose the category 100+, although there are few serious profiles to be found here. Also one-year-olds, toddlers and pre-schoolers are member. It

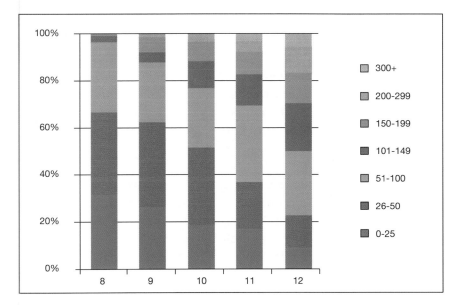

Figure 6.1 • Number of Hyves friends for 8- to 12-year-olds
Source: Mijn Kind Online / Digivaardig&Digibewust / Qrius 2009a

goes without saying that the very youngest only have a digital life thanks to their parents. However, children who can hardly read or write and still have a Hyves page, are no longer an exception. Among the 8- and 9-year-olds approximately one-third can be found on Hyves and for 10-year-olds this is more than 50%. Of the 11-year-olds this is almost two-thirds and of the 12-year-olds three-quarters.[5]

Almost all 8- to 12-year-old members had been member for less than two years in 2009. An explanation for this could be that Hyves had mainly attracted adolescents and young adults (between 16 and 35) until then.[6] Children are not continually online on Hyves, as is sometimes thought. A mere third logs in every day, also one-third 1 to 3 times per week and the remainder even less often. Age does play a role: older children log in more often to Hyves than younger children. Girls also log in more often and, on average, have more Hyves friends than boys.

6.3 • FRIENDS

Adults are often surprised when they see how many 'friends' children and adolescents have on their Hyves. More than 100 friends is no exception, and as the children grow older the number of friends they have also grows (see Figure 6.1).

Children have heaps of Hyves friends, but who are they? Almost 9 out of every 10 questioned children have classmates and their parents in their friend network (see Figure 6.2). The parents of 8- to 12-year-olds are usually still quite young themselves. For these children it is therefore quite normal that their parents are also on Hyves and are also their friend. Approximately one-third of the children even have a grandparent in their network.

The younger children are, the more often they have their father, mother or grandparent as a Hyves friend. At the same time, the younger the children are, the less often they have former classmates, holiday friends and friends from sports clubs as their Hyves friends. This can simply be explained by the fact that they have not yet built up as many contacts as older children have.

There is no difference between boys and girls and the types of friends they have in their network. Girls do have more different kinds of contacts in their list but that is because they log in more often and have more friends.

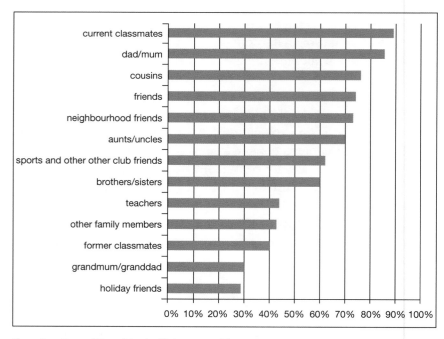

Figure 6.2 • Types of Hyves friends of 8- to 12-year-olds

Source: Mijn Kind Online/Digivaardig & Digibewust / Qrius 2009a

The question, however, remains how children manage to get 100, 200 and even 300 friends. Some of them are probably vague acquaintances, like children in the neighbourhood, children from sports clubs and children whom they have met on holiday (see Figure 6.2). Apart from these, there will also most certainly be 'friends of friends' in the list. These have not been included in Figure 6.2 as they represent a small percentage (23%).

PARENTS AND TEACHERS

Not all children look forward to having parents and teachers as Hyves friends. Hyves pages exist with names like 'Shit my mum's got Hyves' and 'Hate to parents and teachers looking over your shoulder'. Despite this, almost half of the children have a teacher in their Hyves. This gives teachers an extra possibility to keep in touch and show an interest, such as: *"Hi Els! Great that you can type! How are you? How are you enjoying grade 1? I'm fine! Bye! Miss Tina"*.

What do schools and teachers think of this? The opinions appear to be divided. Those against, mention 'fading boundaries', 'too informal' and state that teachers are not meant to be a 'friend'. On the other hand, those for, see no harm and even observe better contact between pupils and their teacher as certain subjects become easier to discuss.

Some schools have rules and protocols, other schools limit themselves to advice (to the teachers – what they should keep in mind), and others do nothing at all. But however well it has been organised (or not): teachers will always have to wonder what it means to have pupils getting a look into their private life.

Discussion in the teachers' Hyves group (more than 12,500 members) on whether to add pupils or not is proof that it is a 'hot' item. Quote: "Not only children but also their parents want to add me. I received several private messages asking how their daughter or son was doing. But that's what the parent-teacher talks are for!" Many teachers appear to like the solution of creating a second Hyves profile so that they can keep work and their personal life separate: *"I've made a Hyves for me as teacher, the kids love it. They scribble a lot and my teacher's email address is also handy (..) It also keeps me up to date with the children!"*

Children can endlessly entertain themselves on Hyves. From scribbling and looking at photos to playing casual games. The 8- to 12-year-olds have indicated which Hyves' functions they consider to be important. A top 20 was compiled from this (see Figure 6.3).

Four groups of activities can be distinguished[7]:
• The first group mainly covers functions focused on communication and contact;
• The second group is focused on gadgets;
• The third group deals with functions related to status;
• The fourth group deals with making friends and new friends.

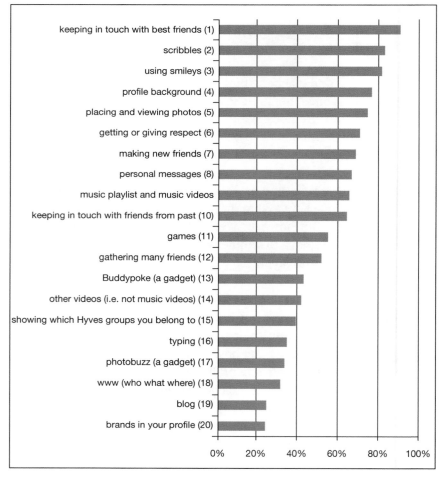

Figure 6.3 • Hyves' functions important to 8- to 12-year-olds
Source: Mijn Kind Online/Digivaardig & Digibewust / Qrius 2009b

Of all the possibilties Hyves has to offer, children indicate that 'contact' is by far the most important. Previous research on Hyves also showed that communication was a much stronger motive to use Hyves for the younger people than for the older Hyves users.[8]

Contact functions include:
• scribbles;
• personal messages;
• keeping in contact with friends (and old friends);
• smiley's;
• getting respect and compliments;
• adding photos and looking at them.

Keeping in contact is what the children consider to be most important (see Figure 6.3 above). Scribbling is also important. As soon as children can type, they start. With touching spelling mistakes they scribble messages to each other, the content of which is often disarmingly simple, for example: *"How sadd for yu that yur granddad is dead!!!!!!!!!!!"* Or: *"Pit that your grandad died, annoying for you, good luck* ☹*"*.

Besides scribbling, it is possible to send someone a private message, which looks very much like a normal email. Children choose for a scribble, private message or use something outside of Hyves depending on the subject. Practically all children ask how people are via Hyves, mainly using a scribble. Most also arrange to do something fun via Hyves, usually as a scribble but also in a private message. About half use Hyves to talk about personal things; this is almost always done via private messages. The other half say they do not use Hyves for this. But this is also done via scribbles, as the scribbles on the death of granddad show. Arguing and gossiping say very little about what children do on Hyves.

Posting photos on Hyves and also looking at photos is also high in the top 20 of favourite activities and also falls under 'communication and contact'. That may appear strange, but other studies have also shown that photos are not only a form of expressing identity but also a means of communication.[9] Children add photos to provoke reactions. Commenting on each other's photos is a way of making contact with the opposite sex. Photos can serve as a starting point for a conversation. They have often creatively edited their photos making them, for example, black and white, merging other images into the photo or increasing the contrast. 'Giving respect' (as the equivalent of Facebook's 'liking' is called) on photos where a lot of work has obviously gone into, can also be a conversation starter. This also applies to *getting* respect. Many children find it nice and important to get respect regarding a photo. It also gives them self-confidence.

The use of smileys is also very important on Hyves. Users usually put them in their scribbles, often very many and also all different. The younger the children, the more important children find smileys. The 8-year-olds even find smileys the most important thing on Hyves.

Keeping in contact with people from the past is particularly important for older children. That is logical, the youngest children have a shorter past.

Girls consider the contact functions of Hyves to be more important than boys do.[10] This was already a known fact: girls are more focused on communication than boys. They also send emails and use MSN more often than boys.[11]

GADGETS

Gadgets – also known as *widgets* – are small applications, mainly focused on entertainment, that can be added to a profile. For example: Buddypoke (avatars), Photobuzz (animation), mini games, horoscopes, digital cats that follow your mouse, rainfall radar and AMBER alert. But also music videos can be seen as gadgets.

Two-thirds of the children find music videos to be important. Roughly half of the children find games such as Buddypoke important. Children can take part in virtual activities together on Buddypoke, portrayed as dolls that they have often adapted to look like themselves. They can cycle together, dance or read books. Unfortunately there are also some less suitable activities for children possible, namely: getting them to trip, laugh at them or kicking them. Particularly for the youngest children this can be harsh as they cannot yet distinguish properly between reality and fantasy.

Adolescents find music videos (and other videos) more important than younger children. On the contrary, Buddypoke and games attract more younger children. Boys find the gadget functions more important than girls. They are particularly interested in games and videos. This is consistent with the information that boys usually play more games than girls (See Chapter 2 – Games).

A user's status includes:
- showing what your favourite brands are;
- showing to what Hyves groups you belong (i.e. your interests);
- letting people know what you are up to.

With regard to the information transfer, this is all one way traffic. It contributes to the image that someone wants to portray of themselves. This looks like spreading an identity but identity actually includes the entire Hyves profile and concerns more than just these functions.

Besides 'friends' and 'privacy', identity is an important aspect of social networks on which much research has been carried out.[12] In adolescence the development of the identity is key and the internet is a good medium to experiment with this. Especially via IM (instant messaging, such as MSN) adolescents present themselves differently for the sake of self-exploration, for example, older, sturdier or prettier – to see how others react. Pre-adolescents (9- to 12-year-olds) actually do this most frequently (see Chapter 5- Online communication).

Experimenting with identies is more difficult on Hyves than on instant messaging (e.g. MSN), because the Hyves profile is actually the digital reflection of a person. So if someone comes across differently on Hyves than they really are, this quickly becomes obvious. But because most friends in someone's network also know the Hyver offline, people are more likely to present themselves in a way received well by friends.[13]

Someone's status can, therefore, be seen as part of the complete digital identity. By putting many brands and expensive brands in their profile, some boys try to hype up their status. They often have really long lists, from Adidas and Calvin Klein to Gucci and Playstation. It does not, however, mean that they actually also own these products.

The older the children are, the more important they find status (on Hyves). This becomes particularly clear at around 13 to 14 years of age. Among the 8- to 12-year-olds there is no difference between boys and girls in the importance of status functions on Hyves.

MAKING FRIENDS

Almost three-quarters of the children find making new friends via Hyves important and more than half wants to collect many friends. There are even Hyves groups aimed at making extra friends. One description of a Hyves group is 'Looking for an extra Hyves friend?': "For everyone who could use an extra friend. For example, to reach the maximum number of 1000 wannabe friends, to feel less lonely, to give yourself the feeling that you are popular."

The following scribble, written by a 12-year-old to a friend, clearly shows the importance of meeting friends: "see dude, you've now got 46 friends and if you hadn't invited all those people you know but didn't want as friend, you would now only have had 20 sooo… I was right". At the same time having too many friends is also not how it should be as you then become unreliable.

As with status, the importance of new friends doesn't reach its top until after the age group on which this research is based, namely, between the age of 12 and 15. This agrees with the results from research carried out in Belgium on profile sites.[14] There is no difference between boys and girls; both find new friends equally important.

As children find collecting and making new friends increasingly important, they also actually have more friends in their network. Of all the 8- to 12-year-olds who took part, one-third has friends in their Hyves network whom they have never seen. These are usually children who have a lot of Hyves friends anyway. Furthermore, as children get older they also have more friends in their network whom they have never seen.

6.5 • SECURITY AND PRIVACY

The security of children on the internet is an issue. It often has to do with people with wrong intentions who try to approach children online. The stereotype men with sweets are no longer just to be found in parks, as they also try to contact children via Hyves. For example: a 35-year-old man from the Dutch village of Nieuw-Vennep approached a 12-year-old boy via Hyves. The boy's mother was looking over her son's shoulder and called in Peter R. de Vries (an investigative journalist) who unmasked the man in a park and handed him over to the police.

Related to security, privacy is also a point of focus. Think of, for example, the photos that children put on Hyves, of themselves in swimwear. Hyvers can therefore choose to protect part or all of their profile. If a profile is protected, only friends in the personal network can see the profile, or possibly friends of friends.

A large majority of the children indicates that their profile is protected. No difference was found between boys an girls. Younger children do, however, have their profiles protected more often than older children. As Hyvers start to find making new friends more important, they protect their profile less often. A disadvantage of putting the profile on the protect mode is that the profiles of others also become invisible. Despite this, a mere one in twenty children gives this as reason to also make their own details visible for everyone. Three-quarters of the children taking part indicated that they do not protect their profile as they have no problems with everyone being able to have access. The remainder said they did not know that details could be protected or did not know how to do this.

It helps if parents set rules for protection. Of the children who say they have no rules for this, 30% has protected their profile. Among those who do have rules, 70% have protected their profile. The majority of all children have made agreements with their parents on this, as is the case regarding with whom they may and may not add to Hyves and how often they are allowed on Hyves.

According to the American researcher Danah Boyd[15] 'talking about it' works better than setting rules. She calls for letting children think for themselves, by asking them questions: "Who do you think will be looking at your profile?" "What do you think is and is not appropriate to put on your profile?" Two-thirds of the children say they talk to their parents about Hyves at least once a month. But: the older the child, the less often this takes place.

One-third says they have also received information on the safe use of Hyves from friends. Children inform each other. Their password, however, is not often passed on to friends but most often to their parents. Of the children up to the age of 10, almost all parents know their password. With increasing age this decreases as does the setting of rules by parents on Hyves and talking about Hyves at home.

6.6 • HYVES VERSUS 'REAL LIFE'

When people talk to each other IRL (in real life) they can see and hear each other. On the phone they do not see each other but can hear each other. When people scribble notes to each other or send emails they can neither see nor hear each other. The receiver misses the facial expression and intonation of the sender. The fewer of these signals, the less 'rich' a communication medium is and the greater the chance that a message is wrongly interpreted.[16] As an 11-year-old girl said[17]: "If someone scribbles *'Did you have a nice holiday?????'*, then you cannot hear whether she is jealous or really interested".

Besides, people are often more direct on Hyves than in real life. They dare to say more (on what they feel or think), or they dare to be unfriendlier than in real life. This is one of the reasons that children sometimes have unpleasant experiences on Hyves: 1 in every 8 children said they had at some stage received an unpleasant scribble or message, making them feel bad. And 1 in every 10 children said they had argued on Hyves. Arguments, however, do not have to have started on Hyves. They can also be continued there after they had been started elsewhere, for example, the school playground.

Another difference between contact via Hyves and contact IRL is that friendships are visible. In real life, friendships can cool off. People see each other less (which is not a bad thing as this is how things go). On Hyves, however, at a certain moment someone is a friend and then suddenly no longer. After removing the friend, the number of friends is clearly one less. At the beginning of 2009, trend watchers forecasted that defriending would be the trend for 2009, although Hyves did not recognize this themselves. More than half of the children say that they have, at some point, removed someone from their Hyves and more then two-thirds have been removed themselves by someone else. In both groups, 2 in every 5 children say that they find this annoying.

A last difference is that children can visit each other as often as they want on Hyves. They can, therefore, also limitlessly compare themselves to the most popular boy or girl in the class. Does he or she have nice photos? More friends? More brands in their profile? Approximately 1 in every 5 children say that they have looked at the profiles of popular children and sometimes become insecure because of the profiles of others. These children find – as to be expected – status and new friends more important than the children to whom this does not apply. They also actually have more strangers on their Hyves.

It is possible that children feel less insecure if they have more friends and are therefore prepared to add strangers. Through status and friends, they may be trying to show that they are somebody or appear to be more popular than they may really be. Furthermore, these children indicate more often that they dare to be more direct on Hyves than in real life. Children who dare to be more direct, and have more strangers on their Hyves – also to be expected – are more often confronted with arguments and undesired scribbles. In addition, children who often look at the sites of others and more popular people and sometimes become insecure because of this, find it more difficult to defriend than children who do not do this. Children differ in their vulnerability on Hyves just as they do in real life.

6.7 • CONCLUSION

Children appear to use Hyves mainly to communicate with each other. They mainly keep in contact with those whom they also see in real life, such as their best friends, via scribbles with smileys. Classmates and parents are the most common contacts that children have in their network of friends. But also grandparents have entered the digital friend network of children.

In addition, children are busy with gadgets on Hyves, such as Buddy-poke or games. They try to increase their status by putting expensive brands on their profile and showing which Hyves groups they belong to. Meeting new friends is also important to a lot of children.

Parents play an important role in security. They set rules on, for example, protecting profiles, and this works. Parents also talk to their children about Hyves, although this becomes less as the children grow older. This is a pity – children become more active on Hyves as they reach adolescence.

Relatively few children have reported problems, such as arguments and undesired scribbles. There is, however, a risk group, namely, children who compare themselves to popular children which makes some of them insecure. They dare to be more direct on Hyves and allow more strangers in to their friend network, which goes hand in hand with more arguments, undesired messages and scribbles.

NOTES

1 boyd, d. & Ellison, N.B. (2007). Social Network sites: Definition, history, and scholarship. *Journal of Computer-Mediated Communication, 13*(1), article 11. http://jcmc.indiana.edu/vol13/issue1/boyd.ellison.html.

2 Qrius is a market research company focusing on young people. The survey was filled in by children from their internet panel.

3 Kol, E. (2008). *Hyves.* Utrecht: Kosmos Publisher.

4 See: www.hyped.nl/details/20090722_raymond_spanjar_over_de_echte_statis-tieken_van_Hyves.

5 These figures have been obtained from the database ASOUK 2008 (Poverty and social exclusion of children) by The Netherlands Institute for Social Research (SCP). This survey asked children whether they have a Hyves profile.

6 Kol (2008). See note 3.

7 Groups formed by factor analysis. Cronbach's α for the groups are, for contact .79 (numbers 1, 2, 3, 4, 5, 8, 10) for gadgets .70 (numbers 9, 11, 13, 14 and 17) for status .72 (numbers 15, 18, 19 and 20) and for new friends .70 (number 7 and 12).

8 Vos, H. (2006). *Networked Individualism: De nieuwe manier van samenzijn* (The new form of togetherness)? A study on the use of social network sites on the internet. PhD thesis Communication Sciences, Amsterdam University.

9 Cleemput, K. van (2008). Self presentation by Flemish adolescents on profile sites. *Tijdschrift Voor Communicatiewetenschap, 36,* 253-69;
   boyd, d. & Heer, J. (2006). Profiles as Conversation: Networked Identity Performance on Friendster. In: *Proceedings of the Hawai'i International Conference on System Sciences (HICSS-39),* Persistent Conversation Track. Kauai, HI: IEEE Computer Society. January 4-7, 2006.

10 In the correlations, crossings were made for each gender with age, how long they have been member and how often they log in and for age with gender, how long they have been member and how often they log in.

11 Duimel, M. & de Haan, J. (2007). *Nieuwe links in het gezin* (New links in the family). The digital world of teenagers and the role of their parents. The Hague: The Netherlands Institute for Social Research (SCP).

12 Peter, J., Valkenburg, P.M. & Fluckiger, C. (2009). Adolescents and social network sites: identity, friendship and privacy. In: S. Livingstone & L. Haddon (eds.) *Kids online. Opportunities and risks for children.* pp83-94. Bristol: Policy Press.

13 boyd, d. (2007a). Why Youth ♥ Social Network Sites: The Role of Networked Publics in Teenage Social Life. *MacArthur Foundation Series on Digital Learning – Youth, Identity and Digital Media Volume.* (ed. David Buckingham). Cambridge, MA: MIT Press.

14 Van Cleemput (2008). See note 9.

15 boyd, d. (2007b). Social Network Sites: Public, Private or What? Knowledge Tree, *13.*

16 Daft, R.L. & Lengel, R.H. (1984). Information richness: a new approach to managerial behavior and organizational design In: Cummings, L.L. & Staw, B.M. (eds.), *Research in Organizational Behavior, 6,* 191-233. Homewood, IL: JAI Press.

17 From a group discussion with 9- to 12-year-olds as preparation for this research.

Bye bye Barbie, hello Stardoll

Lisanne (11) and Nina (10) from Wassenaar shop on the internet everyday to buy designer clothing for their dolls on Stardoll.com. Usually with virtual 'Stardollars', but sometimes also with real money. The nicer your house and your 'me-doll' the more popular you can become.

"Look, you can furnish your house and make a doll that looks like yourself". In Lisanne's room, laptop on her lap, Lisanne and Nina show exactly what Stardoll is. "I look a bit stupid now", giggles Lisanne when she shows her doll, that really is as blond as she is. Her virtual house looks like her real room: the same white wardrobes and even the white grand piano from the living room. Nice things are important on Stardoll because those who have style and 'live' in taste can be chosen as Covergirl, the highest attainable. Lisanne: "Besides, in a nice room you also look nicer yourself."

You pay using Stardollars on Stardoll. You get these free, or you can gain them by playing games. Lisanne knows all the European countries by heart because she is good at the game 'Globetrotter'. "I really didn't know where Portugal was at first and had never heard of Bosnia-Herzegovina, and now I have." But for the really nice things you have to buy Stardollars. With real money. "My mother recently paid 15 euro, I can play for three months with this", says Lisanne. "I didn't even ask for it, she felt sorry for me because there were so many nice things that I couldn't buy." Nina still plays with 'free' dollars: "I often see things that I would like to have but can't buy. That's a pity."

"We're off to go shopping at Starplaza now." Lisanne mentions some brands: "DKNY, Elle, Voile, Vivienne Tam… I had to look the last two up on the internet, I didn't know them. And this is Limited Edition, but that is really expensive. Oh, look, a nice dress, I'm buying it." She drags the dress to her doll and wow, the metamorphosis has taken place. Everything always fits on Stardoll. Nina admits that she'd like to wear some of the clothes herself. But: "Most brands don't have clothing in my size. Although I did recently buy a vest that I'd worn on Stardoll."

Outside, in the hall, there's a long row of forgotten Barbies. Stardoll is much nicer, according to Lisanne en Nina. They admit that they spend more than an hour per day doing it. Often together. Nina: "I don't understand everything yet so Lisanne helps me. And we laugh if we see something really ugly and talk about what we like."

Do you have an idol on Stardoll? Together: "Yes, Lady Gaga!" Lisanne and Nina also want to become famous. Lisanne: "Singer, actress, model, it doesn't really matter which. And if nothing works out, then I suppose that I want to be an architect."

Afke Schaart, Lisanne's mother: "I also go to fashion sites a lot, like Net-a-porter. It's funny to see that Lisanne now recognises things such as the red soles of Louboutin Shoes. Stardoll makes them very fashion-conscious. It's an ideal site for marketing professionals; this is their future clientele. But there isn't much marketing, they don't get any mails with all sorts of offers and can do a lot without having to pay. She is also becoming aware of the fact that you can't buy everything and sometimes have to save. She sometimes looks at something for a month before she buys it.

I think it's nice that they are being creative and that they are developing their own style. It's much nicer than what we used to do with Barbies, much more real and more interactive. I'd really love to do it too, but unfortunately I'm already far too busy."

• www.stardoll.com

7

Advertising

Esther Rozendaal

Ever since children were uncovered as a lucrative marketing target group several decades ago, advertising aimed at children has been a subject of discussion. A key question in this discussion is to what extent children are capable of processing advertisements in a conscious and critical manner.

Globally, two opposing views exist. Many policy makers, parents and consumer organisations are of the opinion that due to lack of knowledge and experience, children are not yet capable of seeing through the temptations of advertising and are therefore more susceptible to advertising effects than adults. This group, therefore, labels advertising aimed at children as being misleading and unfair. The supporters of advertising, often manufacturers and advertisers of children's products, have a different view. They see children as experts and critical consumers.

This discussion has recently once more become the focus of attention on both a scientific and social level, particularly because major changes have taken place in the commercial media environment of children. They are now not only approached more often and at an increasingly young age, but marketing professionals are also making increasing use of new media to promote their products to the young consumer.

The arrival of new media has led to marketing professionals experimenting with new strategies and new marketing techniques with which to reach children. Over the past years, several market research reports have been published giving extensive descriptions of the new types of advertising with which children are confronted.[1] On the basis of these reports, four characteristics typical of advertising in new media can be distinguished:

• product placement;
• targeting;
• user-generated ads;
• viral marketing.

PRODUCT PLACEMENT

Over the past years, we have seen an increase in a combination of advertising and non-commercial content, also called product placement. We know this from television programmes (think of actors eating Lays crisps in a television series, for example), but nowadays brands are also increasingly present in new media, particularly in social media and games. Examples of social media are online social networks (Hyves), virtual worlds (Habbo), instant messaging (MSN) and weblogs.

Product placement is a good way of creating a positive link between consumers and brands. Scientific research among adults has shown that exposure to product placement in both traditional and new media can result in a greater brand awareness and a more positive attitude towards the brand.[2]

A virtual world in which much product placement for children is done is *goSupermodel*, a community where girls make their own virtual model which whom they can communicate, make new friends, play games, chat, exchange photos, share a diary, etc. In virtual worlds such as *goSupermodel*, children are confronted with many banners and billboards but are also directly urged to spend money on virtual products, such as clothing, shoes and accessories to dress their personal avatar.

Product placement is also used in games. You can, for example, stumble across advertising round games, such as banners on game sites, or advertisements shown while loading a game. In addition, advertising is often seen in the game itself, for example, billboards along a racing track or a soccer field. This is also called in-game advertising. Further-

Figure 7.1 • The Superstars of goSupermodel: The girls who score highest on fame.

more, there are also advergames in which the game itself is the advertisement. A product or brand is the focus of advergames. These games are usually free of charge and are distributed via the website of the brand itself (branded space), via the website of another brand (Jetix) or via game portals and other frequently visited sites.

Game portals are websites on which the advergames of various companies can be played. Examples are Spele.nl, FunnyGames.nl and Spelletjes.nl. Often games can be found on these portals aimed at collecting personal details and using these for commercial purposes. Players can, for example, be urged to give the addresses of friends, which are then used to send adverts.

TARGETING

Marketing professionals want their advertising message to be as relevant as possible for those who come into contact with it. One of the ways in which they can best focus their advertising is by targeting, in other words behavioural and profile marketing.

Figure 7.2 • Playing shovelboard with Smoeltjes, an advergame with the leading part played by a biscuit. This children's biscuit is the focus of the site smoeltjes.nl.

In behavioural marketing the online behaviour of children is followed and recorded in a personal behavioural profile, to then be able to approach them with relevant online offers, often as banners. In addition, advertisers also make use of profile marketing, in which advertising takes place based on specific profile characteristics.

Many children and adolescents use one or more social networks where they leave behind a treasure trove of information in their profiles: demographic information, their preferences on certain topics, their friends, hobbies, and the brands they consider to be cool. On the social network site Hyves advertisers can, for example, show their banners on the pages of people who have been selected on the basis of demographic information such as gender, age and address.

USER-GENERATED ADS

Another way of making an advertising message as relevant as possible for the target group is by getting members of this target group to make their own ads. This ties in with the idea that today's generation of children are not merely media consumers but they also enjoy producing media themselves. They are therefore also called *prosumers*.

Marketing professionals make use of this by getting adolescents to create their own advertisements or offering them the possibility of making wallpapers and sending personal e-cards (obviously somehow incorporating the brand).

Apart from allowing companies to make ads with minimal cost in this way, they can also obtain much information on the preferences and interests of their target group. Jumbo, a manufacturer of board games, asked children via the Jetix website to design their own Stratego commercial. Jumbo promised to broadcast the best commercial on Jetix.

In viral marketing or *buzz-marketing* people pass on advertising messages to each other. The internet is a perfect medium to do this: you reach a large number of people in a relatively cheap way.

Viral marketing, is in fact, a modern type of advertising by word of mouth, but much faster and much bigger and therefore more effective. A good example is advergames, in which extra points or an extra chance to play can be obtained if you pass on the game to friends. Films are also suitable for viral distribution. A viral film, for example, a commercial, is produced once after which the distribution is carried out by the target group. Children who like the film will then send it on to friends and acquaintances in their network or put the film on their profile page.

There are two important differences between traditional advertising and new types of advertising:
• new types of advertising are often embedded in entertainment media, such as online social networks and games, making the boundaries between commercial and non-commercial content vague or sometimes even non-existent;
• new types of advertising often do not contain an actual message that children can critically process. Instead, these types are mainly focused on creating positive associations with a brand in a subtle (and often interactive) way. The English scientist Agnes Nairn also calls this evaluative conditioning, or unconscious influencing.[3]

The current generation of children are frequently exposed to advertising in new media. Research on surf behaviour of children shows that the majority of the internet sites often visited contains some form of advertising.[4] But what do we know about how children deal with these new types of advertising? Are they aware that there is advertising in virtual worlds or in games? And do they understand the aims and by whom they are made? In other words, how advertising-literate are children when it comes to advertising in new media?

A fair amount is now known on how advertising-literate children are when it comes to television advertising but little is known on their knowledge and skills when it comes to new types of advertising. Researchers who have shown an interest in this subject over the past years, all seem to agree that the types of advertising found in new media are fundamentally different from traditional types of advertising and therefore form a new challenge for children.

RECOGNISING ADVERTISEMENTS

In order to be able to critically process advertisements, it is initially important that children know that they are dealing with an ad. Only if they can distinguish between what is and what is not an advertising message will they be able to subject this message to a critical view and be able to avert its influences.

Recognising a commercial on television is pretty straightforward for most children. Scientific research shows that children have already fully mastered this skill at around the age of 5.[5] But the fact that children can distinguish between a television programme and a commercial at this age does not mean that they understand its objective. Recognising a commercial is especially based on perceptual characteristics, such as length, speed or the presence of a distinct separation between a programme and a commercial break, and not so much on insight into the nature of the advertising.

Advertising in new media does not, in general, have these recognisable clues thereby possibly resulting in children having more difficulty recognising advertising in new media as such.

Ali, Blades, Oates & Blumberg were pioneers in studying the degree to which children can recognise advertising on websites.[6] They showed

6- to 12-year-olds a number of self-designed websites including one or more blocks of advertising (banners) and then asked them to indicate everything that they thought was advertising. The recognition of banners on websites appeared to grow with an increase in age: 6-year-olds recognised about one-third of the advertising, 8-year-olds about a half and 10- to12-year-olds recognised approximately three-quarters.

Children develop a recognition for banners on websites much later than for television commercials. There are two possible explanations for this.
Firstly, television advertising differs from internet advertising. The main difference is that television commercials are not shown at the same time as a programme while this is the case for internet advertising. In addition, advertisements on websites often contain texts and images that do not differ greatly from the rest of the website. Far fewer perceptual clues can be used to recognise advertising on websites.
Secondly, most children are exposed to television from an early age but do not come into contact with the internet until they are older. Children, therefore, have much more experience with commercials on television than with the banners on websites.

A clue that can help children recognise advertising on websites is price information. The presence of a price does appear to make it easier for children to recognise a banner, but this only applies to older children (10- to 12-year-olds).[7] According to researchers this could result from the fact that older children have a better understanding of the relation between products and prices and that they could have learnt from other media – such as advertising pamphlets and television commercials – that the presence of a price indicates that it concerns advertising.

Research carried out by the My Child Online Foundation (Stichting Mijn Kind Online) has also shown that children have difficulty recognising advertising in new media.[8] By using group discussions and interviews with 8- to 12-year-olds, researchers have concluded that children are reasonably immune to banners, provided that the banners are found on the right hand side of the website.

Children have apparently been taught that many banners can be found on the right hand side of websites and this has become a clue to recognising advertising on websites. This, however, means that some non-commercial content is sometimes wrongly seen as advertising and vice versa. Children mainly appear to recognise banners on the basis of perceptual characteristics, such as location, while they do not yet see these banners as specific advertising messages.

If a website has many images and the banners closely resemble these images, it becomes even more difficult for children to distinguish advertising from the other content. In addition, advertising related to the subject matter of the site is usually not recognised as advertising by children (for example, ads for Disney biscuits on the Donald Duck website or ads for Jungle spread by Zoop on Zoop's website).

It is striking that children do recognise advertisements aimed at adults – e.g. Monsterboard, KLM or Bertolli – as advertising. This changes, however, if the ad offers the possibility to play a game or if the ad is disguised as a game. The banner is then not or scarcely seen as advertising.

UNDERSTANDING ADVERTISING

Being able to recognise an advertising message does not mean that children can see through its tempting nature. In order to be able to do this they have to be capable of looking at the ad from the advertiser's perspective. They have to be able to put themselves in someone else's position, in this case, the maker of the advertising message. According to most cognitive and social development theories, children develop this skill between the ages of 8 and 10.[9] Many changes in how advertising-literate children are, are therefore also expected to occur in this period.

Over the past 20 years, many researchers have been focused on the question to what extent children are capable of understanding the purpose of television advertising. Although they do not all agree as to the exact age at which children start to develop this skill, it can be generally concluded that most children understand at the approximate age of 8 that commercials are not just for fun or intended as handy toilet breaks on television, but that they are meant to sell products.

As children grow older, this insight is further developed and at around the age of 12 most children realise that advertisers are trying to convince them to buy a certain brand or product, by influencing their thoughts and feelings on that brand.[10]

The question is whether this age pattern also applies to the development of children's understanding of advertising in new media. Recent research by the English developmental psychologist Laura Owen shows that this is not the case.[11] Owen studied the understanding 6- to 10-year-olds have of so-called branded websites and ingame advertising, and she compared this with their understanding of television commercials. She showed the children the English website of McDonald's for a few seconds and then asked them if they knew what the website was for. She also showed them a still image of a PS2 game Worms 3D on which one of the worms was holding a Red Bull can. Here too she asked what the purpose of this ad was.

The results of this study show that the understanding of advertising in new media increases with an increase in age but insight into the purpose of branded websites and in-game advertising for all children lags far behind their insight into the purpose of television commercials. Particularly the youngest children (aged 6 and 7) mainly see the new ways of advertising as entertainment while they are capable of seeing the commercial nature of television commercials.

Research reports published by the Canadian Media Awareness Network and the English researchers Fielder, Gardner, Nairn and Pitt show similar results for the understanding children have of advergames.[12] Both reports conclude, on the basis of interviews, that most children – and even teenagers – who play advergames see them as fun games and not as advertising.

The Australian scientists Mallinckrodt and Mizerski also carried out research on the understanding children have of advergames.[13] They had 5- to 8-year-old children play a Kellogg's advergame for five minutes. Points can be won in the game by getting Toucan Sam to throw as many Froot Loops and pieces of fruit in a monster's mouth as possible. Obviously, the Kellogg's Froot Loops logo can be seen clearly throughout the duration of the game. After playing, the children were asked what the purpose of the game was. Approximately half of the children realised that the game wanted them to buy and eat Froot Loops. They were also asked whether they knew who had put the game on the internet. Only a quarter of all the children was able to identify Kelloggs as the source of the game. An explanation for this limited knowledge is that the children in this study did not yet have sufficient experience with advergames resulting in them not being able to recognise the commercial nature of this new type of advertising.

The American communications expert Ellen Seiter also concluded that the insight of children into the source of advertising in new media is poor. She talked to 8- to 12-year-olds about Neopets, a virtual world popular in both America and the Netherlands, and chokerblock with advertising. The children were not aware of the commercial nature of the site and all thought that it does not cost any money to make a world like Neopets and put it on the internet. In answer to the question why they think Neopets exists, most children said "Neopets is there because someone somewhere had a good idea and put it on the web for our entertainment". [14] They see the maker of Neopets purely as someone who enjoyed doing it, more or less as a hobby.

Conclusion: 6- to 12-year-olds have difficulty seeing through the commercial nature of advertising in new media. An explanation for this is that an entanglement of commerce and entertainment have made it difficult to distinguish the advertising message from other content and to identify the source of the message.

7.3 • ADVERTISING EDUCATION AND ADVERTISING MARKERS
One of the most frequently discussed ways of improving advertising literacy among children is advertising education. Advertising education aims to make children aware of the commercial nature of advertising and the strategies used by advertisers to sell their products but also to develop a critical attitude among children to advertising.

In the Netherlands and many other Western countries several teaching tools and other educational services on advertising are now available, although they mainly deal with traditional types of advertising (See the Mediawijsheidkaart.nl for an overview of what is available in the Netherlands).

There are indications that advertising education can improve children's insight in traditional types of advertising but if this also applies to new types of advertising had, until recently, never been studied.

This lack of knowledge inspired the American communications expert Susannah Stern to determine whether advertising education, in this case a short group lesson on advergames, could improve children's understanding as to the commercial nature of these games. [15] In order to study this, she asked 9- and 10-year-olds to play a Kraft Foods adver-game for 10 minutes. The results show that children who had had the lesson on advergames before playing had a better understanding of the fact that the purpose of the game was to get to like Honey-Comb break-fast cereal and they could more easily identify the source of this adver-game. However, although they were able to identify the source more easily, their understanding remained poor.

Stern also carried out research into other ways of making children aware of the commercial nature of advergames, namely, the presence of clear markers, such as the word 'Advertisement'. It is generally thought that such markers can be a clue for children to activate their awareness. It, however, does not appear to work. The indication 'Adver-tisement: this game and other activities on this website contain mes-ages on products sold by Kraft' does not help children to identify the source or see through its purpose.

A possible explanation is that children simply do not see the marker. The markers used by advertisers are generally subtle or placed in an inconspicuous part of the website thereby easily missed. In addition, it is also possible that the meaning of the marker is not clear because of use of difficult language. In order for advertising markers to work, vis-ibility (size, colour, location) and clarity (child-friendly language) of the markers are vital. Advertising markers on websites and in adver-games aimed at children could even be standardised so as to make it even easier for children to learn to recognise advertising on websites and in advergames.

7.4 • DOES ADVERTISING LITERACY MAKE CHILDREN RESILIENT?
Are children who can see through the commercial nature of advertising also less susceptible to its effects? This question has, despite its rele-vance, scarcely been studied. The limited number of studies carried out come to the unexpected and maybe even disturbing conclusion that this is not the case.

The Australian scientists Mallinckrodt and Mizerski found that the in-sight of children into the purpose of the *Froot Loops* advergame had no influence on their preference for *Froot Loops* compared to other cereal brands or other food categories such as hamburger, sandwiches and fruit.[16] Children who were aware of the fact that the advergame was intended to sell Froot Loops were influenced by the game to the same degree as children who did not realise this.

This is in contrast to what you would expect on the basis of the assumption that knowledge on the commercial nature of advertising makes children more resilient to its influences.

The Dutch communication experts Rozendaal, Buijzen and Valkenburg also came to the conclusion that advertising literacy does not by definition make children more resilient to the influences of advertising.[17] They discovered that being aware that the aim of television commercials is to make people like the products being advertised, meant that for certain product groups with much advertising (sweets, toys and computer games), 10- to 12-year-olds bought less of these products. But: for younger children their desire for these products only increased.

There are a number of important reasons to doubt the notion that knowledge on the commercial nature of advertising makes children more resilient to its effect.

Firstly, it is very likely that children do not use this knowledge when they process advertisements, even though they have familiarised themselves with it.

Secondly, the effect of advertising is to a large extent, and for advertising in new media probably almost entirely, determined by emotional reactions. In other words, a child can still be tempted by advertising even though he or she has the necessary advertising knowledge. Future research will have to reveal whether and how the advertising literacy of children is related to their susceptibility to advertising effects.

In conclusion, it is fair to say that children are a unique consumer group and the specific characteristics of this target group will have to be taken into account when developing and regulating advertising directed at them.

NOTES

1 Among other.: Calvert, S. (2008). Children as consumers: advertising and marketing. *The Future of Children, 18,* 205-234;
Fielder, A., Gardner, W., Nairn, A. & Pitt, J. (2007). *Fair game? Assessing commercial activity on children's favourite websites and online environments.* National Consumer Council of the United Kingdom. [Electronic version]. Downloaded 23 July,
2009 from http://www.agnesnairn.co.uk/policy_reports/fair_game_final.pdf;
Reklame Rakkers (2008). *Nieuwe vormen van reclame anno 2009* (New forms of
advertising in the year 2006). [Electronic version]. Downloaded 6 July, 2008 from
http://www.reklamerakkers.nl/upload/1221400827.pdf;
Mijn Kind Online (2008a). *Gratis! (maar niet heus). Dossier over digitale reclame
voor kinderen.* (Free ! (but not really). File on digital advertising for children).
[Electronic version]. Downloaded 6 July, 2008 from http://www.mijnkindonline.
nl/dosier_digitale_marketing_kinderen.html.

2 Auty, S. & Lewis, C. (2004). Exploring children's choice: The reminder effect of
product placement. *Psychology & Marketing, 21,* 697 – 713;
Lee, M. & Faber, R.J. (2007). Effects of product placement in on-line games on
brand memory: A perspective of the Limited-Capacity Model of Attention. *Journal
of Advertising, 36(4),* 75-90;
Van Reijmersdal, E.A. (2007). *Audience reactions toward the intertwining of adver-
tisement and editorial content.* Thesis Amsterdam University: The Amsterdam
School of Communications Research (ASCoR).

3 Nairn, A. & Fine, C. (2008). Who's messing with my mind? The implications of du-
al-process models for the ethics of advertising to children. *International Journal of
Advertising, 27,* 447-470.

4 See Fiedler et al (2007). See note 1.
Mijn Kind Online (2008b). *Klik en klaar. Een onderzoek naar surfgedrag en usability
bij kinderen.* [(Click and go. A study into surfing behaviour and usability among
children). Electronic version]. Downloaded on 23 July, 2009 from http://www.
mijnkindonline.nl/download/MKO_usability_rapport_download_full.pdf.

5 For an overview, see Rozendaal, E., Buijzen, M. & Valkenburg, P.M. (2009). Do
children's cognitive advertising defenses reduce their desire for advertised products? *Communications, 34,* 287-303.

6 Ali, M., Blades, M. Oates, C. & Blumberg, F. (2009). Young children's ability to recognize advertisements in web page designs. *British Journal of Developmental Psychology, 27,* 71-83.

7 Ali, Blades, Oates & Blumberg (2009). See note 6.

8 Mijn Kind Online (2008b). See note 4.

9 John, D.R. (1999). Consumer socialization of children: A retrospective look at
twenty-five years of research. *Journal of Consumer Research, 26,* 183-213.

10 Rozendaal, E., Buijzen, M. & Valkenburg, P.M. (2008). Reclamewijsheid in ontwikkeling. Een vergelijking van de cognitieve reclamevaardigheden van kinderen
en volwassenen. (Becoming media-wise. A comparison of cognitive advertising
competencies of children and adults). *Tijdschrift voor Communicatiewetenschap,
36,* 270-283.

11 Owen, L., Lewis, C., Auty, S. & Buijzen, M. (2009). Is children's understanding of
non-spot advertising comparable to their understanding of television advertising?
*Paper presented at the 59th annual conference of the International Communication
Association,* Chicago, 21-25 May.

12 Media Awareness Network (2005). Young Canadians in a Wired World Phase II Key Findings. Ontario, Canada: Media Awareness Network. [Electronic version]. Downloaded 23 July, 2009 from http://www.media-awareness.ca/english/research/YCWW/phaseII/;
and Fielder et al (2007). See note 1.

13 Mallinckrodt, V. & Mizerski, D. (2007). The effects of playing an advergame on young children's perceptions, preferences, and requests. *Journal of Advertising, 36*(2), 87-100.

14 Seiter, E. (2005). The Internet Playground. In J. Goldstein, D. Buckingham and G. Brongere (ed.) *Toys, Games, and Media,* pp. 93-107. Mahwah, NJ: Lawrence Erlbaum. Quote on page 100.

15 Stern, S.R. (2009). Increasing children's understanding of advergames' commercial nature: Does an advertising literacy lesson of ad brake make a difference? *Paper presented at the 59ᵗʰ annual conference of the International Communication Association,* Chicago, 21-25 May.

16 Mallinckrodt & Mizerski (2007). See note 13.

17 Rozendaal, Buijzen & Valkenburg (2009). See note 5.

"Everyone in my class has one"

Katinka (9) and Sarah (9) are best friends and both have, despite their young age, mobile phones. They had to do quite a bit of lobbying though.

Katinka: "I really had to nag for my phone, because I'd wanted a phone for a while. Everyone in my class has one. When I turned nine, I was finally allowed one. Now I can phone home to ask if I can leave the after school care. I used to have to ask if I could phone, that was a bother."

Sarah: "I've also got one since my birthday. I was very happy! I can now ask Katinka if she wants to come and play or wants to go on Hyves. I also text message a lot and have 15 great games: Snakes, Powerball, Pokemon etc… And I listen to music, to nice ring tones. It's a prepaid Nokia phone and I get the credit from my dad. I still have 20 euro which will last me a while. It's a normal phone. Katinka has a very nice one, a sliding one."

Katinka: "Yes, I have a Samsung sliding phone. My dad got discount on it. I also listen to a lot of music: Amy MacDonald, Michael Jackson and the Red Hot Chili Peppers… and I play games."

Sarah: "I'm never allowed to turn it on at the table, my parents find this unsociable. And at night I put it on my desk next to my bed. I also use it as alarm clock"

Katinka: "I'm also not allowed to use it during dinner, I put it on the cupboard. At night I put it in my mum and dad's drawer. I'm allowed to take it to school but have to turn it off in the classroom. The teacher gets angry if it rings or if you get a text message."

Sarah: "If the teacher finds out, you have to hand it in and don't get it back till the end of the day."

Katinka: "I'm not just allowed to phone anyone, I have to ask first. But in emergencies I can, of course, phone. If something happens and my parents can't be reached, then I can phone my grandparents."

Sarah: "Yes, I was ill recently and my granddad picked me up at school. He wanted to phone my dad but didn't have his number. I did have it, so I could help him."

Stephan Massalt, Katinka's father: "We have a rule at home that we let our phones be, as much as possible. I think it is important to pay attention to the things at home and I want to show Katinka that these devices don't dominate everything."

"She did indeed nag a long time for it. She had some good arguments: that she had to cycle by herself to gymnastics and that she had to be able to phone us if something was wrong. We decided to get her one and thought her birthday was the right moment."

"Before she got her phone I explained that prepaid meant that she didn't have unlimited use and that she should mainly call or text about important things. But she can phone a friend to ask her to come and play."

"She does spend a lot of time using her phone. She phones, sends text messages, listens to music…she grabs her phone as soon as she has nothing to do. Although she just as easily throws it in the corner when she goes out to play."

8

Mobile phones

Marion Duimel

Owning a mobile phone has become the most normal thing in the western world in 2010. Children, too, own a mobile phone more and more frequently. This makes it easier for them to keep in contact with their parents, keeping them informed about their whereabouts. Parents can also more easily contact their children to tell them they have to come home. Children can of course also text message their friends, to find out, for example, where they are and what they are doing. Apart from that, mobile phones today are also a source of entertainment as you can take photos, make films and play games on them.

What do children use their mobile phone for the most? For entertainment, contact with friends or to be reachable for their parents? Mobile phones are available in all sorts and sizes. From glittery pink phones to models that are mainly meant for music or photos. What is important to children? What rules do parents set and are they related to their own mobile phone use? Do children with parents who use their mobile phone often also use their phone more often or in a different way?

These questions will be answered below on the basis of a survey carried out in August 2009 among 691 Dutch children aged between 8 and 12.[1]

8.1 • MOBILE PHONE, YES OR NO?

The mobile phones marketed at the end of the 90s literally date back to the last century. Looking at these large models with bulging antennas and small screens it almost evokes sentimental feelings. But still it is no more than 10 years ago that people actually walked around with these phones. Mobile telephony is one of the fastest growing technologies and has in approximately 10 years developed from a technology for the happy few to an essential technology for everyone.[2]

With an ever increasing number of people owning a mobile phone, the number of children with a phone is also growing. Half of the respondents had a mobile phone in 2009.[3] There is a big difference with regard to age. Of the 8- and 9-year-olds approximately one quarter had a phone and for the 10- and 11-year-olds this is more than half. Among the 12-year-olds (who were looked at separately because this is the age at which most Dutch children go to secondary school) the majority of children had a mobile phone. Only 2 out of every 10 have to do without.

Practically all children without a mobile indicate that their parents feel that they do not need one yet. In a clarification many children say that their parents think they are still too young or say that they will get one when they go to secondary school. A few children themselves felt that they do not need one yet. Money appears to play a minor role. Less than 1% says that their parents think it is too expensive and less than 1% also think it is too expensive because they have to pay it themselves.

Whether a child does or does not have a mobile phone is closely related to whether someone in the child's direct environment has one. The family plays the key role. Of the children without a mobile, almost two-thirds say that none or only a few family members have one. Of the children with a mobile this is only 1 in every 10. Secondly, the number of children in class with a mobile is also a good indicator. Only 1 in every 6 children without a mobile says that in their class approximately half or more children have a mobile. Of the children with a mobile this is as many as 4 in every 5.

8.2 • FIRST MOBILE

"When you go to secondary school you get a new bicycle, I can remember it exactly, the shop door, the smell in the shop and there it was all shiny" sings the famous Dutch entertainer Youp van 't Hek in his song 'My first bicycle'. Nowadays, many children also get their first mobile phone when they go to secondary school, or even earlier. They often take it with them for safety reasons, whether or not on their new or second hand bike. A 12-year-old Dutch girl writes: *"because I have to cycle through the polder and can phone if something happens"* and another 12-year-old girl: *"in case I have a flat tyre at secondary school."* But also for children still at primary school safety is the main reason. *"Mummy thought it was safer for when I play outside"*, writes a 10-year-old girl. Sometimes a specific event leads the parents to provide their children with a cell phone: *"because we had lost my sister at the fair"* (boy, 11 years old). Almost all 8- to 12-year-olds with a mobile phone say they were given one to be easily reachable by their parents.

A first phone can be obtained in a special way: *"got it from my grandma who passed away"*, writes an 11-year-old girl. An 8-year-old boy *"found one on the street."* Children also indicate that the phone itself is also important: *"because they are cool"* and *"because they are really hot"* (two girls aged 10). A 12-year-old boy gave a very special reason: *"My hearing is not good. Being able to communicate through texting is great for me."*

When comparing subscription models and prepaid accounts, prepaid mobile phones are used most (4 out of every 5 children). An advantage of prepaid cards is that the child cannot spend more than the credit they have on their card. The degree to which the child's parents use their own mobile phone has no consequences for the kind of subscription their child has. Children of parents who use their mobile phone frequently do not have a subscription more often than prepaid compared to children of parents who do not use their mobile phone often. Children do have a subscription more often as they grow older and surprisingly girls also have a subscription more often than boys (22% compared to 13%).

8.3 • PHONING OR TEXT MESSAGING

The two most important functions of a mobile phone for children are making phone calls and text messaging. As children grow older, there is a very clear shift in the relation between calling and texting (see Figure

8.1). Of the 8- and 9-year-olds, only half use text messaging. Children in this age group use their mobile mainly to call their parents. The 10- to 12-year-olds phone their parents even more frequently but the increase in texting with their parents is remarkable. The older a child, the more often it also texts its parents in addition to calling.

A second shift, with regard to age, is related to the contact with parents compared to contact with friends. As children grow older they have less contact with their parents and more contact with friends.[4] This can be seen clearly in their phoning and texting behaviour. The 12-year-olds text their friends more often rather than calling them. An 11-year-old girl says: "I think it's pretty cool to text and then know where my friends are hanging out."

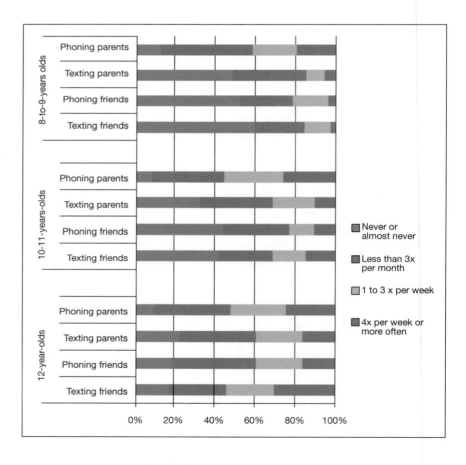

Figure 8.1 • Phoning versus texting with friends and parents among 8- to 12-year-olds.
Source: Mijn Kind Online / Digivaardig&Digibewust / Qrius 2009b

Finally, it is remarkable that as children grow older, the relation between phoning and texting between parents and children is more equally divided. Younger children are more frequently called or text messaged by their parents than they themselves contact their parents. This difference decreases among 10- and 11-year-olds. The 12-year-olds phone and text their parents as often as their parents phone and text them.

It appears that it is generally the mother who contacts the child by mobile phone. The relation between the mobile phone use of the mother and the frequency with which the child is called is much stronger than with the mobile phone use of the father. Also, when stating the reason for having a mobile phone, the mother is mentioned twice as often as the father.

Some topics are most suitable for texting, others for phoning and some topics are not at all suitable to be discussed by mobile phone. Two-thirds of the children use their mobile to arrange to do something fun with friends. More children in this group say they do this by calling rather than texting. A majority use their mobile to ask how friends are – usually by texting rather than calling. Almost one-third say they use their mobile for a personal chat, mainly by calling. Gossiping or arguing is scarcely done by phone. Asking someone out or breaking up is only done by 5% of the children but when it is done, it is usually done by texting. Older children and girls use their mobile phone more often to arrange to do something than younger children and boys (it has been taken into account that older children and girls call and text more often anyway). Girls use their mobile phone slightly more often than boys do to ask how someone is.

8.4 • OTHER APPLICATIONS

Nowadays, children can do a lot more with their mobile phones than just phoning and texting. Figure 8.2 shows the top 10 mobile phone functions for 8- to 12-year-olds. Phoning and texting are number 1 and 2 and taking photos is found at number 3. Two-thirds occasionally take photos with their mobile and almost half even does this every week.

Favourite subjects for talking photos with mobile phones are: family and friends. In addition, children also like taking photos of themselves, of animals and of nature. Or, as a 9-year-old girl wrote: "something

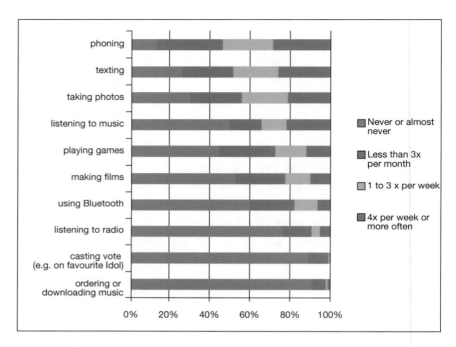

Figure 8.2 • Mobile phone applications among 8- to 12-year-olds.
Source: Mijn Kind Online / Digivaardig&Digibewust / Qrius 2009b

crazy I just happen to see." And: "almost everything I like very much"
(boy, 11 years old).

Children mainly show their photos to friends. A mere 1 in every 5 put
their pictures on the internet, e.g. on a Hyves (Dutch social Network)
page. Some children use a photo as background on their phone, others
save these photos (on their mobile phone or computer) and have a look
at them once in a while. With a mobile phone in their pockets, children
constantly have the opportunity to be creative by taking photos and
making films. Some use this opportunity regularly, such as this 12-year-
old boy: *"I enjoy showing photos to my friends and making videos with
friends for YouTube."*

But there are even more possibilities. Approximately half of the children use their mobile to: listen to music, play games and make films.
Quite a few children also use Bluetooth with which they can send photos, etc. to each other, and 1 in every 10 children say they cast votes with
the mobile, e.g. for programmes like Dutch Idols. Merely 1 in every 20
of the children order a ringtone, game or wallpaper for their mobile

(this is why it cannot be found in the top 10 of Figure 8.2). Also surfing, MSN, Hyves or watching TV via the mobile phone is only done by 1 in every 20 children, and then sporadically.

Most parents probably think their children are still too young for mobile internet, which is relatively expensive. Considering almost all children have their phone specifically for safety reasons, their parents probably think internet is unnecessary. Enjoying yourself with your phone by taking photos and making films, playing games and listening to music usually does not cost anything, in contrast to other internet functionalities.

Texting, taking photos, making films as well as voting by phone are done more often by girls than boys. Older children use their mobile more often to take photos, listen to music and Bluetooth than younger children. Children also phone and text more often as they grow older, although this is not too bad. Children do not phone or text message all the time. About half of the children use their mobile about three times per month to do this or sometimes never at all. Finally, if children do not text message it is simply because they do not like doing it, because they find it difficult or because they simply do not know how. An 11-year-old boy writes: I can't text, it doesn't really interest me." And a nine-year-old boy: "texting annoys me and phoning is much easier, my mum put the numbers in for me."

Parents' mobile phones are also regularly used by children. Half of the children say they sometimes phone using their mother's or father's phone. Children who do not have their own mobile do this more frequently than children who do have one (68% against 41%). A quarter of the children text using their parents' phone. In this case it is irrelevant whether they have a phone themselves or not. A third of the children take photos and play games with their parents' phone. In addition, 1 in every 10 use their parents' phone to vote, e.g. on a TV programme. Children who do not have their own phone do this more often than children who do. Some mobile phone-less children are sometimes allowed to use their parents' phone to do fun things. But also children who have their own phone can borrow one from their parents. These children may not have a camera on their phone or their father or mother has a more advanced phone and other games.

There is a strong relationship in the use of various functions of the mobile phone. Four groups can be distinguished, namely[5]:

- The first group has to do with primary functions: phoning and texting;
- The second group deals with entertainment, e.g. games, photos, movies, music and Bluetooth. (children send each other photos and videos they have made via Bluetooth);
- The third group is about 'ordering' (ringtones, games, wallpapers), downloading music, voting and buying things in a virtual world (e.g. furniture for Habbo Hotel);
- The fourth group is about functions that require an internet connection. E.g. surfing, watching TV, using MSN and Hyves and using Google maps or GPS.

  Children who make frequent use of one specific function within one particular group (e.g. taking photos) are also highly likely to use the other functions within that group (e.g. playing games).

8.5 • WISHES AND EXPECTATIONS

What do children expect from a mobile? More than 9 in every 10 consider phoning to be important and three-quarters think that the ability to text message and take photos is important. Approximately half find the possibility to play games and listen to music important. A 12-year-old boy writes: *"It's like a computer, also fun to do. Playing a game or making a film is easy in the car. Then it's not so boring."*

A mobile should not just be able to do things, it must look good too. Approximately 3 in every 5 children consider a nice phone to be important, as this 12-year-old boy does: *"it's cool to have a nice mobile."* And a 10-year-old girl writes: *"it's cool to have a mobile especially if your friends don't have one."* About 1 in every 5 children want the latest phone. The fact that they are expensive only concerns 1 in every 10.

Older children and girls find it more important that their mobile looks nice, that they can text and take photos with it, than younger children and boys. Particularly texting is more often 'very important' for girls than for boys (42% against 26%). Furthermore, the older they get the more important it is to be able to listen to music.

Some characteristics are strongly related to each other. Three groups can be distinguished:[6]
- **functional use** – phoning an texting;
- **status** – nice, expensive and 'the latest'. Children for whom this is important often talk about their phone as being 'cool', 'hot' and 'neat';

- **entertainment** – e.g.: playing games, listening to music and being able to take photos.

In practice, it goes without saying, children do not only find a phone important for its functionality, or as a status symbol or for entertainment. All these aspects can, however, play a role, although the use of the mobile may be restricted by parents, as this boy's (8) statement shows "it looks cool, taking photos is really nice but I'm usually not allowed to use it - only if we go somewhere, an amusement park or something." The surveyed children gave other reasons to explain why they think a mobile is important. Quite a few children wrote that they get a feeling of 'belonging' if they have a mobile. A boy (9): "almost everyone has a mobile phone, it is really cool, if you don't have 1 it's stupid."

Children also like mobile phones because it makes them feel 'big'. As if they fit in more with grown-ups. The privacy aspect also plays a role, talking about private things with friends in their own room. But nowhere near all children find a mobile important. Plenty of children, in fact, do not like their mobile at all, like this boy (9): "I don't really like it but at least I can now let my parents know where I am or if I'm late my parents can phone me too."

An 11-year-old girl: "I don't like it but it is important to be able to phone my parents if I need to." Also this boy (9) clearly says that only functionality is important for him: "My parents made me get a phone so that they can reach me when I'm playing outside. I never call anyone and have had my first prepaid for six months now." That a feeling of safety can change into a feeling of being restricted is shown by the following boy's (12) statement: "I think they only want to know everything. How much I spend and check up on you." This boy only has his phone in case of bike failure and is not allowed to do anything else with it.

8.6 • THE ROLE OF PARENTS

Most parents have given their children a mobile with a view to safety and being reachable. In most cases they have made agreements on it's use but not always. Below is an overview of the findings, including a division of the costs.

For one in every 5 children no rules have been set as to the use of their mobile phones. Parents who use their phone intensively do not set rules more or less often than parents who hardly use their phone. The phone behaviour of parents itself does not appear to be related to setting phone rules for their children.

If rules have been set they are often (for 6 out of every 10 children) related to the money they are allowed to spend on phoning and texting. In 1 out of every 10 children the phone behaviour is restricted to a limit on the time and frequency the child is allowed to call and text. The older the child, the more often this agreement is made. These rules also apply more often to girls than to boys. This is not strange, particularly with regard to text rules as girls text more often than boys. A 12-year-old girl tells how her mother keeps and eye on her: "I enjoy texting my friends. Sometimes I do it too often because I get a lot of messages from friends. I often get a warning from my mum for this." It makes no difference whether a child has a subscription or a prepaid; rules restricting the use apply equally often to both groups.

For slightly less than half of the children, agreements have been made at home on whom they may call and text. This rule applies more often to younger children (almost two-thirds) than to older children (more than a quarter). It is likely that young children are still only allowed to phone their parents. This is shown by their phone behaviour (see Figure 8.1).

One-third of the children have to turn off their phone at night and a third also have agreed with their parents on other places where they have to switch off their phones, e.g. on their bike or at a party. Both rules apply as often to boys as to girls, while age is also irrelevant.

BEARING THE COSTS

A large majority of parents pay for the costs of their children's mobile phones. They do this more often – understandably so – for younger children. For 9 in every 10 children with a subscription, the parents pay the bill. In the remaining cases, half pay their own bill and the other half split the bill with their parents. For children with a prepaid phone, parents pay the bill in 8 out of every 10 cases. In the remaining cases, half pay their own bill and the other half split the bill with their parents. The National Institute for Family Finance Information (Nibud) would like to see more children being given the responsibility for their own bill so they can learn to manage a budget.[7]

Children with a subscription spend more money on a monthly basis than children with a prepaid phone. Of the children with a prepaid phone, almost three-quarters think they spend between €0 and €10 a month, of the children with a subscription it is more than half. Children with a subscription often think that the amount is higher and they often also do not know how high the costs are of children with a prepaid phone. In any case, there are hardly any children who think the costs exceed €20. To save costs, children can also phone their parents and get them to call them back. A boy (11) said: *"it's fun and easy. Phone mum and mum phones me back."*

8.7 • UNDESIRABLE SUBSCRIPTIONS

The Opta (Independent Post and Telecommunications Authority of the Netherlands) regularly gets complaints of undesirable and unintentionally expensive text message services to which you are suddenly stuck. This can happen after ordering a – so-called free – ringtone, a game or a wallpaper or after having played a text message game on TV. Advertising for said ringtones are often seen on music channels. More than half of the children say they see an ad on TV for a ringtone, game or text message game at least once a week. But this is also the case on internet. One in every 8 children with a mobile phone is asked for their mobile number on a weekly basis – or more often – via internet, e.g. while playing a game. Children who have been asked for their number more often say they have had an undesirable subscription more often. A total of 1 in every 10 children with a mobile phone has had an undesirable subscription. These are – as to be expected – particularly the children who have at some point ordered a ringtone or voted on a TV programme.

There is no relation between the rules set by parents (e.g. on whom they are allowed to text or how much they can spend on phoning and texting) and undesirable subscriptions. It is of course possible that children applied for a service by mistake while their parents had given them permission to order a ringtone or vote once. A subscription linked to something can also be unclear to adults, in part due to the notorious 'small print'.

An increasing number of young children have a mobile phone. They have practically all been given the phone with safety and being reachable in mind. The younger the children, the more often the phone is only used for contact with their parents. Texting and contact with friends increases as children grow older. This is probably why, in contrast to reports that adolescents are bulk consumers of texting, phoning is also the most important function for this age group. It is conceivable that this is different for slightly older children.

Apart from a functional use (phoning and texting) the mobile phone is suited for the entertainment of children, e.g. listening to music, playing games. In addition, children can also use their phone for creative purposes by taking photos and making videos. Only very few children order products such as ringtones and games and also very few children use their mobile to vote. Despite this, 1 in every 10 children say they have had an undesirable text messaging subscription. Very few children also use their mobile phone to go on the internet.

Some children who do not have a mobile themselves, are allowed to use their parents' phone. Children who do have their own phone do, however, also use their parents' phone, although this percentage is lower than that for phone-less children. Parents play a major part in the finances; the majority pay for the user costs. Most parents also set rules on the use of the mobile phone. First of all, there are agreements on how much a child can spend on phoning and texting and secondly, whom the child is allowed to phone and text.

Children find it particularly important that they can phone and text, but entertainment features (photos, music and games) are also high on their wish list. Most children find the appearance of their phone important but it does not have to be particularly expensive or brand new. The last two characteristics only appeal to a minority. There is also a group of children not in the least interested in a mobile phone. For this age group, who simply assume mobile phones have always been around, it is certainly not a 'must have'. In any case, safety comes top as the reason to get a mobile phone. All the same, this does not mean that the majority do not have lots of fun with the mobile phone.

NOTES

1 The survey was commissioned by the Stichting Mijn Kind Online (My Child Online Foundation) and the joint venture Digivaardig & Digibewust (Digi skilled & Digi aware) and carried out by market research company Qrius. Qrius is a research agency which focuses on youth. The survey was completed by children in their internet panel.

2 Castells, M., Fernandez-Ardevol, M., Linchuan Qiu, J. & Sey, A. (2007). *Mobile Communication and Society: A Global Perspective.* Cambridge, MA: MIT Press.

3 These figures on ownership were taken from the Qrius youth study 2009, which is carried out every two years, among approximately 1000 8-12-year-olds.

4 Larson, R. & Richards, M.H. (1991). Daily companionship in late childhood and early adolescence: Changing developmental contexts. *Child Development*, 62, pp. 284-300.

5 Groups were formed by factor analysis. Cronbach's α for the groups are for functional .75, for entertainment .81, for ordering .89 and for internet .94.

6 Groups were formed by factor analysis. Cronbach's α for the groups are for status .83 and for entertainment .72.

7 See: http://www.nibud.nl/over-het-nibud/actueel/nieuws/nieuws/artikel/nibud-geeft-ouders-onvoldoende-voor-financiele-opvoeding-1.html.

"Isn't it great that the whole world can see what we're doing"

Puck Meerburg (10) and Nando Bennis (11) won a Microsoft design competition for their website 'Blackspots'. Children can calculate the safest route to and from school on the site. Besides being good friends they are also ideal partners: Puck is programmer, Nando is the creative one.

Puck: "We won an Xbox 360 in Microsoft's competition. And they liked our website so much that we were also allowed to present it at a conference."

Nando: "An American agent of Microsoft came to us there and gave us his card. He was amazed that we could do this and said that he wanted to help us if and when we had new plans. That was great to hear."

Puck: "We now also have a company – Coolkit – so that we can make websites for people. I'm proud of this."

Nando: "Puck is a real programmer and I'm more business-like, and like doing things. Thinking up a plan, pitching and then making it, that's what I like the most. I design it using Adobe Illustrator and Puck works it up. I often sleep over at his place and we often come up with good ideas. I had already been building websites for two years but when I met Puck we could do it together. That was great. Most other children don't understand what I do. The nice thing about making websites is that everyone around the world can see it."

Puck: "I think I probably use the computer more than the children in my class. But, no, I'm not bullied because of this. In fact, they ask if I can help them with their computer problems, even the principal asked for my help once."

Nando: "I also enjoy doing other things on the computer. I'm now learning to play guitar using the programme Guitar Band. And I also play outside. During a cycling holiday in Denmark I didn't see a computer for weeks. Oh yes, and I have a fossil collection. Yes, I like the newest period and things that are millions of years old."

Erik Meerburg, Puck's father: "I think many children have a certain focus, some know a lot about dinosaurs and others about trains. Puck didn't get his interest for computers and networks from us, although he does see us use the internet. He is self-taught and is strongly driven to sort things out in great detail. Puck and Nando made Blackspots themselves, I really couldn't do that. Sometimes he goes a bit too far and then I say: 'go and do something else for a while'. On the other hand, he doesn't watch TV and I think this is more interactive."

"Puck sometimes really talks like an adult programmer but in other ways he's just a kid. This sometimes leads to an unbalanced situation. His classmates do not always understand him. There was a direct click with Nando when they met, through us; Nando's mother and I are colleagues. They immediately sat down behind the computer together. They phone and email each other almost every day. This friendship is very valuable. Without Nando, Puck would only be able to talk about IT with adults."

• www.puckipedia.nl/blackspots
• www.coolkit.nl

9

Information skills

Els Kuiper

New names are continuously being invented for new internet generations. In 1998, Don Tapscott talked about the *Net Generation¹,* Marc
Prensky in 2001 about *digital natives²*, and in 2006, Boschma and Groen
about the *Generation Einstein³*. The purpose is the same each time: a
matter of a gap between younger an older generations with regard to
the use of ICT in general and the internet in particular. Many parents
and teachers see this gap daily; children appear to be smarter in their
use of ICT than adults and they are also very self-assured in this area.
The media gladly picks this view up and strengthens the suggestion that
children learn to use all types of ICT applications automatically.

But how skilled are children really? Teachers in primary and secondary
education regularly doubt the value of using the internet as a source of
information. They wonder what children actually 'learn' from it. Their
pupils proudly present professionally looking school projects and give
talks using flashy Powerpoint presentations. But their papers have ex-
tracts which the teacher doubts the pupils understand. It is also unclear
from which websites the information was obtained, let alone whether
the pupils paid attention to the usefulness and reliability of the information.

This chapter deals with the use of the internet as a source of information for education purposes. Information skills are part of the broader concept 'media literacy'. Being able to critically make use of digital information is an important characteristic in a society increasingly characterised by large amounts of information in which users have to be capable of finding their way. Education can fulfil a role in acquiring information skills.

In this chapter we look at what scientific research can teach us on the use of the internet as a source of information for children and the skills they do or not possess. We also look at the possible differences between groups of pupils and the consequences for the upbringing and education. The chapter concludes with an analysis of the content and design of websites.

9.1 • INFORMATION VIA THE INTERNET

People have always needed specific skills to be able to manage information. But the arrival of the internet has greatly changed the nature and the amount of information. Children are now growing up in a society in which an increasing amount of information is available digitally. This calls for *new literacies*: new types of literacy, partly related to reading in 'old' sources of information such as newspapers and libraries, but at the same time covering new and different skills that are necessary in using digital sources of information such as the internet.[4] Which skills are involved here?

The internet is vast in size and does not have an unambiguous structure as is the case in printed sources of information. This means that it is, first of all, important to know what you are actually looking for. Formulating a clear question is, therefore, even more important when using the internet than it is in the use of traditional sources of information. This does not merely concern school related questions. When you are looking for information on the internet you are always looking for something, whether it be the recipe for apple pie or measures against the greenhouse effect. These questions can also be broadly formulated: e.g. you are looking for everything about Michael Jackson. But even in this case the internet is question driven. You should be able to hold on to this question during the search. This prevents the search for information becoming a goal in itself.

Being able to localise the information then becomes a skill that is much more important on the internet than it is when using traditional sources of information. The latter, after all, usually have a clear structure: a book has page numbers, chapters and sometimes an index. Internet information does not have such a linear build up. All the information can be found more or less parallel to each other and is connected to each other through links. There is therefore not a previously determined way to manoeuvre through the information: every internet user chooses his or her own route by continually having to make decisions on whether to click on links or not. That starts with a Google search: which results on your screen do you click on first and what do you consider? An internet user more or less constructs his or her own 'text' by going from website to website. In traditional or offline texts that process is determined more by the author of the text.

In localising information it is, therefore, not just skills directly related to the *search* for information that play a role. Being able to *read* and *understand* the internet information is also of vital importance.

Reading on the internet is not the same as reading books. Children themselves also see it as something different: they often don't even realise they are reading (or having to read) on the internet. Despite this, reading skills are very important on the internet partly because much information on the internet was not written for children. This includes, in the first place, all non-school related information, i.e. everything children search in the areas of sports, music, hobbies, games, etc. But even when children are looking for information for school there is a large chance they come across all sorts of things written for adults. Internet asks its users for a flexible use of reading strategies, e.g. skimming or more precise reading.[5] School reading lessons usually pay little attention to online reading.

Internet is also very accessible: everyone can leave all kinds of information. This is very appealing to children; they can, after all, also easily become authors of internet information. But on the other hand, they can also easily be confronted with incomplete, wrong and sometimes potentially harmful information. An important skill when using the internet is therefore being able to *assess* whether the information is understandable, useable and reliable. In assessing the reliability of sources on the internet prior knowledge on the subject plays an important part: the more prior knowledge available the easier it is to properly assess a source. Children, therefore, often lag far behind as they have less prior knowledge than adults.

Finally, the internet demands skills with regard to synthesising, or in other words, putting the information together in order to come to a coherent answer. Particularly because users follow their own path in their search for information it is very important to be able to compare and connect the information.

Figure 9.1 • Screenshot Google. The search engine Google is popular among children, irrespective of their age.

International research has now given us some details on the way in which children search for information on the internet. [6] What picture does this give?

To start with, children gladly use search engines like Google, irrespective of their age. Very young children, who cannot yet read or write well, already make use of the possibility to search for images with Google. From about the age of 8, Google is the most obvious way to access information for many children. Recent Dutch research has shown that almost 90% of 12-year-olds use Google to search for information on the internet. For 8-year-olds this is almost 50%.[7] The fact that Google is becoming increasingly user-friendly helps, for example, the question "Did you mean...?", if a word is not spelled correctly.

Children's search engines, in which children only have access to previously selected websites that correspond with their age, only appear to be appealing for young children. From the age of 9, children clearly see the restrictions of this kind of search: you can usually only find school-type information.[8]

Coming up with good search words can be difficult for children. For example, they use complete sentences as search term and do not always know the tricks such as the use of inverted commas. Besides, they make a lot of spelling mistakes; younger children as a result of their limited language ability and older children as a result of sloppiness and haste.[9] Spelling mistakes, however, sometimes have no consequences because of the correction, mentioned earlier, offered by Google.

In addition, children often try URL's. If you are looking for information on Michael Jackson, you simply type www.michaeljackson.com and wait for the result.[10] Slightly older children (from about 10 years of age) also try other extensions, e.g. .com or .org. Children also prefer to use familiar websites. If they have found good information on a country on the Dutch website www.landenweb.nl, they will use this again next time they want to know something about a particular country.

Children often simply click on the first result if they search using Google.[11] They do look at other results but do not really read them. This means they could miss important information. They often know how websites work: they recognise menus and links and they know how they can be used. However, in practice, they often do not use them; they quickly zap through a page and miss important links or on the other hand they hang around very long in a certain text. They also often appear to really trust Google: it 'knows what they are looking for'. This can result in remaining in the same search strategy too long; they are convinced it must be 'somewhere'.

Assessing the internet information is very difficult for children.[12] They often know the difference between the internet and a book and that internet has information on it that is untrue. But they do not know how they can judge this. Many children are critical of advertising and therefore disregard websites with many advertisements.[13] They do not, however, always know what is and what is not advertising. A website with a menu structure and many coloured blocks would then, for example, not be used. They also do not always realise that commercial websites can also yield useful information. You can, for example, make good use of the McDonald's website if you are looking for information on the composition of McDonald's products. More subtle ways of influencing are even more difficult to see through, e.g. the websites of political parties and charities.[14]

For children of primary school age the synthesis of information they find is still very tricky. They do often know that you have to look at different places for information and that you then have to make your 'own text' but in practice they quickly get bogged down by the enormous amounts of information that they find. The fact that they are inclined to look for a ready-made answer to their question also plays a role. The following example is a good illustration of this.[15]

Two girls are looking for the answer to an investigative question they thought up themselves: "What is the difference between free-range and battery eggs?" They find a website in which is explained, in two text boxes, what free-range and battery eggs are. They read the boxes and then click the website away. When asked why they could not use this website they reply: "It does say what free-range eggs and battery eggs are but it does not give the difference."

Good use of the internet in actual fact calls for research skills, which include deciding on your question, localising, assessing and synthesising of information. This is something children do not automatically learn at school or at home. Education is often still directed at finding 'the only correct answer'. If children are used to this you cannot expect them to approach the internet in a more investigative manner.[16]

Research also shows that children's achievements vary greatly when they look for information on the internet for an investigative or other question.[17] Particularly children aged around 10 do often have a command of the necessary skills, such as formulating search terms and the skimming of web texts. They do not, however, use these skills consistently. General achievement level does not appear to affect this. Good reading skills also do not directly result in more successful internet use. It appears that there are underlying skills playing a role not directly related to the information skills of children.

It is important to have a certain degree of *flexibility* when searching on the internet. Always adhering to the same strategy is not a good idea; alternating strategies and realising on time when something is 'not working' is more useful. Many children have difficulties with this. In contrast to what parents think, they explore remarkably little on the internet and often keep going round in circles.[18]

A substantial share of *patience* is also indispensible on the internet: finding good information often takes longer than expected. However, most children have a shorter span of attention than adults. They also expect to quickly find answers: that is why internet is so appealing.[19] The tendency for children to look for ready-made answers and thereby fully trust Google can also be counterproductive.

Effective and efficient internet use, in fact, comes down to a combination of skills and *reflection ability.* The latter means: keeping an eye on what you are doing and why, and adjusting in time if your strategy does not yield enough.

Particularly young children do not yet have this ability to reflect on their actions. Until the age of about 10 they are 'concrete thinkers' and have difficulty with the abstraction necessary for such reflection. In addition, something paradoxical seems to happen when using the internet: because children expect that everything can be found ready-made, they appear to be less inclined to actively and critically gather knowledge themselves, while the internet calls upon such an active and critical attitude, more so than the traditional sources of information.[20]

The above particularly shows the skills that children lack when they use the internet as a source of information. This is, however, not the complete picture. Children are very handy with 'buttons'. They possess, almost naturally, many instrumental skills that are also necessary when using the internet.[21] More so than adults, they intuitively feel how you can navigate within one website and between websites. This applies to more than just the internet; many parents recognise the situation in which they try in vain to tune the DVD recorder with the help of the instructions manual and their child then effortlessly solves the problem by pressing a few buttons.

These instrumental skills are an expression of the self-evident manner in which many children go about with the computer and the internet: not as something difficult requiring many skills but something that you can just 'do'. Children make a lot of use of the trial and error method. In other words: simply trying something. And if it does not work, you try something else.[22] In many situations this can be a good method but it cannot be compared to consciously and reflectively searching for information. This shows that the information skills that children are expected to use at school, can be opposed to their own – spontaneous – manner of working. The latter, however, is also of value.

Recent English research on the way in which very young children (3 to 6 years old) use text on the computer, shows that children who cannot yet read are still capable of dealing with textual instructions on a computer screen.[23] The children were able to indicate that the button with the word *play* meant that they could play a game and the button *exit* meant that they could stop the game. In other words: the children gave a meaning to a word in the same way as to a picture or icon despite the fact that they had no idea what it really said. Trial and error was clearly an important strategy; children tried out what the word on which they clicked meant.

Over the past years, there has been a call, from various different directions, for another style of education doing more justice to the way in which the current generation of pupils 'learn'.[24] This involves intuitive and non-linear build up of knowledge, which is not sufficiently covered by traditional text books and classroom education. The question, though, is whether this plea sufficiently takes into account the differences between pupils. However, it can do no harm to realise that children partly use other (and by definition not less valuable) strategies than their parents and teachers, e.g. trial and error. This gives information on the way in which children can best acquire information skills. For example: not from a book but learning by doing, making room for trying and making mistakes. It also lays the foundation for continuing dialogue. The relation between teachers and pupils is, after all, very different when using ICT than when teaching reading or maths, in which case the pupil is always the one who does not know something yet and the teacher helps the pupil further.

It is also good to take into account that many children use the internet outside of school for many different applications and certainly not always only to look for information. It can be worthwhile to connect the internet use at school, and the demands made of the pupils there, to pupils' internet use in their leisure time, thereby acknowledging what they 'learn' in this way.

9.4 • DIFFERENCES BETWEEN GROUPS

It is important to recognise that children do not form one homogenous group. They can differ in gender, ethnic background, socio-economic situation and achievement level. Not much is known yet on the influence of these differences on the internet skills of children under the age of 12. Differences between boys and girls appear to play a much less important role than for more technical computer skills: girls are slower and more precise, boys faster and less precise.[25] Both strategies can be more or less successful depending on the context. Non-native children can be less language proficient than native pupils but it is unclear to what extent this influences their information skills. Particularly informative websites can often be very language rich which can be difficult for all weak readers. Differences in achievement level do not necessarily translate into differences in information skills; qualities mentioned earlier, such as flexibility and patience, also play a major role.[26]

There has been renewed attention in recent years for digital inequality, in other words: the differences in possibilities of using digital media. Initially, the main focus was on differences in computer ownership and access to computers whereas now the focus is more on differences in skills. Terms such as media literacy also play a role. The thought is that managing media and digital media is a condition for citizens to be able to function in society.[27] Both international and Dutch research shows that as far as this is concerned it is a matter of inequality between groups of children and adolescents, where the socio-economic background plays a role.[28]

For nowhere near all children is the use of computers and the internet self-evident. It is therefore not a question of a uniform generation of digital natives; children differ greatly in the appreciation of and the manner in which they deal with computers and the internet. In a recent study on the degree to which students of the Dutch preparatory secondary vocational education (VMBO) resemble the image sketched of the 'Generation Einstein', this is also confirmed.[29] The conclusion is that VMBO students in general and non-native VMBO students in particular, have far fewer of the skills to be able to make use of all the opportunities offered by society; information skills are part of this.

It is to be expected that these differences already play a role in primary education. There lies a great challenge for education: how can acquiring critical and reflective internet use take shape for all groups of pupils? How can it be prevented that certain groups can make less use of the opportunities of the digital society because of a lack of skills? Recognition of the existence of these different groups of pupils with a variety of characteristics, skills and possibilities is the initial step to be taken.

9.5 • WEBSITE DESIGN

To what extent does this subject also involve designers of websites and digital material? Children can, after all, have difficulties with websites with a not easily accessible and very wordy structure. Recent research has shown that children between the ages of 7 and 9 were able to remember information better if a website used a branched menu structure with images and symbols rather than a list with just words.[30] For children aged between 10 and 12 this made no difference. In addition, the use of learning aids, such as repeating the most important information the moment a child wanted to go to another web page, proved to be functional.

Dutch research also shows that the appearance of a website can be of influence on the way in which children deal with it.[31] It was mentioned earlier that children are very alert with regard to advertising and that they click away anything that looks like an ad. Websites with normal images are then unjustly clicked away. At the same time they are not capable of seeing the difference between advertising and editorial content. For the designers of websites there is much to be improved, although mixing advertising with information is of course often knowingly done. Designers of informative websites for children should, in any case, take the skills children do and do not have more into account.

'Pimping' with the use of images is not always necessary; more important is a clear navigation structure, clear text boxes with a not too small type face and the functional use of illustrations. The makers do not have to worry that they will come across as being too informative; American research has shown that secondary school students can easily distinguish between websites that are meant to be informative and are suitable for school work and websites that they can use in their leisure time. Students expect informative websites to contain large amounts of text.[32]

Taking everything into account it is good to be aware that children make use of the entire internet from a young age. For school work they may limit themselves to websites suitable for children but at home they do not. At home they look for information on hobbies, sports and music and also make use of many more and different internet applications than at school. Adapting websites and making them more child friendly is therefore desirable but not sufficient to compensate for the differences in skills (and the possible lack thereof). Particularly schools have an important task in this. It would, therefore, be beneficial if teaching guidelines for 'information skills' were to be developed to give children the opportunity to learn to critically use the internet as an information source, from a young age.

NOTES

1   Tapscott, R. (1998). *Growing up digital.* The rise of the Net Generation. New York: McGraw-Hill.

2   Prensky, M. (2001). Digital Natives, Digital Immigrants. *On The Horizon* - The Strategic Planning Resource for Education Professionals, 9 (5), pp.1-6.

3   Boschma, J. & Groen, I. (2006). *Generatie Einstein: slimmer, sneller, socialer.* Communiceren met jongeren van de 21e eeuw (Generation Einstein: smarter, faster, more sociable. Communicating with youngsters of the 21st century). Amsterdam: Pearson Education.

4   Coiro, J., Knobel, M., Lankshear, C. & Leu, D.J. (2008). Central issues in new literacies and new literacies research. In J. Coiro, M. Knobel, C. Lankshear & D.J. Leu (eds.), *Handbook of Research on New Literacies* (pp. 1-23). New York/London: Lawrence Erlbaum Ass.

5   Sutherland-Smith W. (2002). Weaving the literacy Web: Changes in reading from page to screen. *The Reading Teacher*, 55(7), pp.662-669.

6   Research into the use of internet as information source by children is usually small scale and of a qualitative nature. It is therefore hardly possible to present the results in concrete figures. The power of the research results lie mainly in their repetition. Most of the research is aimed at children aged between 8 and 12. There are data available on children under the age of 8, but they are often taken from educational practice, not scientific research.

7   Mijn Kind Online (2008). *Klik en Klaar. Een onderzoek naar surfgedrag en usability bij kinderen* (Click and ready. A study into surfing behaviour and usability among children). [Electronic version]. Downloaded 25-06-2009 from http://www.mijn-kindonline.nl/1552/surfgedrag-kinderen-onderzocht.htm.

8   Kuiper, E. (2007). *Teaching Web literacy in primary education.* PhD thesis. Amsterdam: Free University.

9   Mijn Kind Online (2008). See note 7;
    Kuiper, E., Volman, M. & Terwel, J. (2005). The Web as an information resource in K-12 education: strategies for supporting students in searching and processing information. *Review of Educational Research, 75*(3), pp.285-328.

10 Kuiper, E., Volman, M. & Terwel, J. (2008). Students' use of Web literacy skills and strategies: Searching, reading and evaluating Web information. *Information Research*, 13(3) paper 351. [Available at http://InformationR.net/ir/13-3/paper351.html].

11 Kuiper et al. (2008). See note 10.

12 Mijn Kind Online (2008). See note 7;
    Kuiper et al. (2008). See note 10;
    Pritchard A. & Cartwright, V. (2004). Transforming that they read: helping eleven-year-olds engage with Internet information. *Literacy*, 38(1), pp. 26-31.

13 Mijn Kind Online (2008). See note 7.

14 Kuiper (2007). See note 8.

15 Kuiper, E. (2007). *Wat weten we over... webwijsheid in het PO en VO?* (What do we know about...being web-wise in primary and secondary education?) Zoetermeer: Stichting Kennisnet.

16 Loveless, A., DeVoogd, G.L. & Bohlin, R.M. (2001). Something old, something new... Is pedagogy affected by ICT? In A. Loveless & V. Ellis (eds.), *ICT, Pedagogy and the Curriculum: subject to change* (pp. 63-83). London: Routledge Falmer.

17 Kuiper et al. (2005). See note 9;
    Walraven, A., Brand-Gruwel, S. & Boshuizen, H. (2008). Information problem solving: a review of problems students encounter and instructional solutions. *Computers in Human Behavior*, 24, pp. 623-648.
18 Kuiper et al. (2008). See note 10.
19 Mijn Kind Online (2008). See note 7;
    Kuiper et al. (2008). See note 10.
20 Kuiper (2005). See note 9.
21 Ten Brummelhuis, A. (2006). Aansluiting onderwijs en digitale generatie (Connection between education and the digital generation). In J. de Haan & C. van 't Hof (ed.), *Jaarboek ICT en samenleving 2006. De digitale generatie* (Yearbook ICT and society 2006), pp.125-141. Amsterdam: Boom.
22 Walraven et al. (2008). See note 17.
23 Levy, R. (2009). 'You have to understand words... but not read them': young children becoming readers in a digital age. *Journal of Research in Reading*, 32(1), 75-91.
24 For example Veen, W. & Vrakking, B. (2006). *Homo Zappiens. Growing up in a digital age.* London: Network Continuum Education.
25 Large, A., Beheshti, J. & Rahman, T. (2002). Gender differences in collaborative Web searching behaviour: An elementary school study. *Information Processing and Management*, 38, 427-443.
26 Kuiper et al. (2008). See note 10.
27 Raad voor Cultuur (2005). *Mediawijsheid. De ontwikkeling van nieuw burgerschap.* (Council for Culture. Media literacy. The development of new citizenship) The Hague: Raad voor Cultuur.
28 Peter, J. & Valkenburg, P. (2006). Adolescents' internet use: testing the "disappearing digital divide" versus the "emerging digital differentiation" approach. *Poetics*, 34, pp.293-305;
    Facer, K. & Furlong, R. (2001). Beyond the myth of the 'cyberkid': Young people at the margins of the information revolution. *Journal of Youth Studies*, 4(4), pp. 451-469.
29 Groeneveld, M.J. & Van Steensel, K. (2008). Kenmerkend vmbo. Een vergelijkend onderzoek naar de kenmerken van vmbo-leerlingen en de generatie Einstein. (Typically vmbo. A comparative study into the characteristics of vmbo-students and generation Einstein). Hilversum: Hiteq.
30 Rose, M., Rose, G.M. & Blodgett, J.G. (2009). The effects of interface design and age on children's information processing of Web sites. *Psychology & Marketing*, 26(1), pp.1-21.
31 Mijn Kind Online (2008). See note 7.
32 Agosto, D.E. (2002). A model of young people's decision-making in using the Web. *Library & Information Science Research*, 24, pp.311-341.

"Our computer assignments are really quite boring"

Louella (10) and Amy (10) are in grade 5 at a protestant primary school. They think their school should spend more time on computer lessons. "Wouldn't it be great if a computer screen just popped up out of every desk?."

Louella: "We have computers at school but they're in the corridor. Last year I had a few computer assignments but I haven't had any yet this year."

Amy: "I'm in another class and we have three computer assignments per week that we do on our laptops in the class. We do spelling and learning to tell time, for example."

Louella: "I thought the computer assignments that we got were pretty boring. They try to make it fun – we had to build a wall within a certain time for maths for example. But I'd prefer to see a zoo with a speech balloon with a sum in it: an x number of animal dies, an x number is born, how many are left? I'd also like to learn how you can draw using Paint as I always find this very difficult."

Amy: "I think we should be allowed on the internet for, say, half an hour. We're never allowed to. And digital blackboards would be great too."

Louella: "I've never heard of that, we have normal chalk blackboards."

Amy: "Wouldn't it be just great if all the desks had a small computer with a screen that pops up? If it's not too expensive of course. Then we don't have to keep walking to the computer. And then you can look up a word without having to get a dictionary."

Louella: "I'm quite good with the computer. I used to spend much time on the computer but now I'm only allowed for one and a half hours per day. Then I go on Hyves, or play a game or I play on the Wii."

Amy: "I hardly ever use the computer at home. But when I do I go to Spele.nl and Spelen.nl. Or I play Runescape, I have to defeat monsters. And I write stories on the computer, romantic ones or just nice ones."

Louella: "Am I better than the teachers on the computer? I don't know about my female teacher, but my other teacher says he doesn't know much about computers himself."

Amy: "My teacher helps me so she knows quite a lot."

Sjoerd Jaasma, teacher grade 6: "I think we spend enough time at school on computer lessons. Every class has at least three laptops or five computers on the corridor that they can use. Each child should be able to spend thirty minutes per week on the computer. Although a separate computer classroom where an entire class could work at the same time would be ideal of course."
"My sixth graders do spelling exercises on the computer and school projects. Or I get them to google something, State Opening of Parliament, for example. They are not usually allowed to use the internet, we have filtered that. Although they often find out how they can get round that."
"We will soon be getting a digital blackboard. It will be put in our empty classroom. They cost 5,000 euro each, I think, but we do hope to have one in each classroom within a year. These blackboards have great advantages. You don't have to write or draw difficult sums for Maths on the board every year, you just pull them out of last year's archive. And for biology you can easily just show a DVD. That's great."
"Whether pupils get lessons on the internet is actually determined by each individual teacher. We should I suppose set a standard for everyone. We do sometimes get people calling each other names because of an argument that started at home on MSN. We then talk about this."

10

Education

Alfons ten Brummelhuis

Since the rise of new media, some 30 years ago, a long hard road has been followed to come to an effective use of ICT for education purposes. The acquisition of materials, such as computers, digital blackboards and internet connection was successful. The integration of these facilities into the educational curriculum and the actual lessons showed much more resistance.

Effective use of new media in education is a problem many schools struggle with on a daily basis. Below we sketch the present state of affairs, focusing on what new media have brought to education.

10.1 • MISCONCEPTIONS

There are different misconceptions on the role and significance of new media in education.[1] One of them is that children would no longer need to be taught how to deal with ICT. In other words: their computer skills are overestimated. Children who have been brought up with internet are generally sufficiently ICT skilled but they often lack the skills necessary to be able to learn with the help of ICT.[2] Education can only benefit from the wealth of information offered by internet if children have a command of the so-called information skills (searching, selecting and assessing information). Most children have a poor command of these skills and they do not obtain them automatically even though they grow up in an environment rich in new media.

A second misconception is that the current generation of children is better at multitasking than previous generations. The use of several media simultaneously (listening to music and sending emails; watching TV and reading the paper) is also commonplace to a great extent among adults and is by no means new. In the past multitasking also took place, e.g. listening to the radio and at the same time cleaning the vegetables or darning socks. The question now presented in education is what the significance of this multitasking ability of children is. Multitasking can lead to a loss of concentration and cognitive strain when the brain has to shift between competing stimuli.[3] With the exception of talented pupils, multitasking in education means a chance of reduced learning achievements for many pupils.

A third misconception is that the current education system does not fit in with the way in which modern children learn and that educational reform is necessary for the internet generation. It is clear that not all members of the internet generation are the same. The claim that an entire generation has comprehensive knowledge and skills with regard to new technology (the digital natives) is not, in any case, tenable on the basis of empirical research. Equally untenable is that 'they' have their own learning style. The call for educational reform on the basis of these arguments must be answered with reserve. This does not, however, mean that there are no other reasons for educational reform. Below three reasons are stated.[4]

10.2 • THREE REASONS FOR MODERNISING EDUCATION

The availability and development in the area of new media provide three arguments for the modernisation of education. These are:
• Learning objectives;
• Practical training;
• Quality improvement.

NEW MEDIA AS LEARNING OBJECTIVE

The first reason for educational reform is that it is the duty of the education system to prepare pupils for their life as a citizen in a media rich knowledge society. This requires education to include new media in their curriculum. In the past, this subject matter was referred to as 'citizen informatics'[5] or 'IT studies'[6]. Nowadays, there is a lot of talk in the Netherlands about being 'media-wise'[7] and having 'information skills'[8]. The contribution of education to the preparation of children in their functioning as citizens in an ICT rich society is internationally increasingly referred to as *21st century skills*.[9]

The different terms that point to the social preparation of children, have in common that one single definition is lacking. This puts education in the difficult position of having to prepare children for a society which does not yet exist or for which the training requirements are unclear. There is, however, an international development that information skills are, in future societies, expected to play the role of a new type of literacy which every citizen will have to have a command of.[10] This means that information skills will, in future, be as vital as reading, writing and maths.

This position is currently not taken by information skills in education. In the Netherlands, schools are free to choose the focus given to information skills. In practice, this means that schools and teachers can determine for themselves how much attention they give to information skills. This also applies to other subjects preparing children to function in an ICT rich society, e.g. safe internet use and social values in ICT use, such as netiquette, bullying and copyright.

The digital literacy essential for everyone in the future, becomes dependent on the situation at home and the education possibly on offer. The lack of learning objectives in the educational system for learning how to handle new media could result in a divide that forms a threat to the social ambition of preparing every citizen for participation in the information society.

NEW MEDIA IN PRACTICAL TRAINING

The second reason for modernising education with the help of ICT is preparing pupils for the use of ICT applications necessary in their later profession. This necessity, usually secured in professionally qualifying learning objectives, mainly applies to vocational education.

NEW MEDIA AS A TOOL FOR QUALITY IMPROVEMENT

And thirdly: new media can make education more appealing, more effective and more efficient, which results in quality improvement. Not using these opportunities would mean insufficiently using pupils' talents and causing the costs of education to be higher than necessary.

This third rationale is further substantiated later in this chapter. Can education really be improved and made less expensive by using new media? The following questions are discussed consecutively:
• Which new media have set in in primary education? (10.3)
• How are they used? (10.4)
• What do pupils learn from using new media in lessons? (10.5)
• What barriers have to be conquered? (10.6)
• What do these insights signify for the future structure of education? (10.7)

10.3 DIGITAL INFRASTRUCTURE

Primary schools in the Netherlands have, over the past years, made major investments in acquiring computer facilities. In 1998, primary schools had, on average, 1 computer for 27 pupils. The availability of computers then increased in the period 1998-2003 to 1 computer for every 7 pupils. Since 2003, the number of computers in primary education has been stable: on average, 1 computer for every 6 to 7 pupils (see Figure 10.1). Approximately 10% of these computers is a laptop.

MODERNISATION

Schools have been investing mainly in modernisation over the past years. Approximately 20% of the computers are being replaced annually by more modern hardware. In addition, most primary schools have

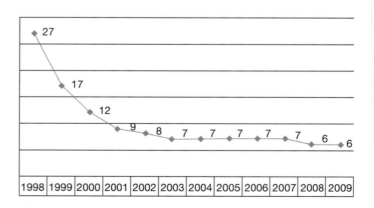

Figure 10.1 • Number of pupils per computer in primary education.[11]

also, in a short space of time, acquired digital blackboards. As from 2006, more than half of the primary schools switched from chalk blackboards to digital blackboards in a period of 2 years. It is expected that in 2011 just about every primary school will have digital blackboards (see Figure 10.2). This means that after the introduction of the computer (mid 1990s) the current advance of the digital blackboard can be considered to be the second digital revolution in education.

INTERNET

At present, almost all computers in primary education (93%) are connected to internet. Approximately one quarter of the schools have wireless internet (28%) and/or an optic fibre connection (26%). This is, however, much less than in other educational sectors. In secondary education more than three-quarters of the schools have an optic fibre connection and more than two-thirds have wireless internet.

The limited availability of fast internet connections at primary schools is also expressed in the degree to which they make use of applications requiring these connections, e.g. webcams and video conferencing. Only 1 in every 10 primary schools use webcams or video conferencing. It is also expected that a mere 3 to 4% of the schools will deploy webcams and video conferencing in the coming year. This means that the majority of the schools will not be making use of real time communication or synchronous image communication, for the time being.

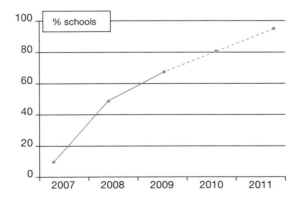

Figure 10.2 • Dutch Primary schools with at least 1 digital blackboard (dotted line is forecast according to school managers).[12]

How are new media used in education? We discuss this question from two perspectives: from that of the school and from that of the pupils.

SCHOOL

Nine out of every 10 teachers in this country's primary education use computers in their lessons. And in the near future almost every teacher in primary education will make use of ICT applications, with or without a digital blackboard.

The most frequently used applications are exercise programmes and games. Many games have an educational purpose but they are also used as entertainment and as a reward for smart pupils. In addition, pupils use word processing and electronic learning environments.

Internet is mainly used for finding information. The frequency of using internet by pupils is restricted and can be typified as incidental rather than regular.[13] Serious games and digital tests belong to the least used applications.

If pupils use the computer at home for school work then it is usually to find information on the internet. 'Looking for information' is the most important school activity for which primary school pupils use the internet. Approximately half of the pupils in grade 5 and 6 feel that they get good tips at school for looking for information. As the pupils grow older, and go to secondary education, their satisfaction on instructions given at school on looking for information keeps going down. In the first stage of secondary education 43% of the pupils are satisfied with the support offered by teachers while in the 3rd and 4th years this percentage drops to 30%.[14]

Almost half of the pupils in primary education is satisfied with the degree to which they come into contact with digital teaching materials at school. According to 3% of the pupils the use of new media is too much while 44% feel it is too little; 8% has no opinion.

Particularly the advance of the digital blackboard has resulted in an increasing number of teachers making use of video fragments in class. At present, two-thirds of the teachers use video in their lessons every week. They expect the use of video to increase further in the near future: 85% of the teachers intend to make use of video every week with-

in two years, of which 60% every day.[15]
In total, teachers use computers almost 8 hours per week and they expect this to increase by more than 40% in the coming 3 years. On the
basis of this estimate, pupils who go to school in 2012 will receive about
11 hours of teaching per week supported by ICT applications. This
would – according to teachers – also mean that the limit for the amount
of time that is effective has been reached. Teachers in primary education say this limit lies at around 40% of the teaching time.

The introduction and advance of ICT applications has not led to a radi-
cal turnaround in the design and organisation of learning processes but
has manifested itself as a gradual development that slowly results in
another education system. The use of ICT is not reserved for a specific
educational vision. ICT is used in various pedagogical approaches, varying from knowledge transfer (teacher decides on exact teaching programme) to knowledge construction (children themselves partly design their learning process).

For many teachers knowledge transfer plays a greater role than knowledge construction. Also in the future this will remain important. But at
the same time teachers would like to teach more often in accordance
with knowledge construction. According to teachers and management,
ICT is an important tool to facilitate this change in the education system. A stage has now been reached in which computers have become
an integral part of education.

Teachers and school managers feel that this integration is nowhere near
complete and they expect that the role of ICT in education will increase
sharply over the next years. This growth will take place on three fronts:
• The number of teachers making use of computers;
• The number of hours that ICT applications are used;
• The number of different ICT applications.

A growing *number* of teachers and pupils are making *more use* of ICT for
learning objectives and in an *increasing number of different ways*. This
means that ICT is not only a replacement of parts within the existing
curriculum but at the same time it is enabling the transformation of
education through the implementation of innovative education pack-
ages that would have been impossible without ICT. (Also see: 10.5 -
Yields).

An average primary school class has pupils with varying experience with new media. Four types can be distinguished:
• **traditionalists** – dominant in primary education (52%), are pupils who make use of the basic, older internet facilities (email, web surfing, information search and MSN). This group is characterised as 'traditionalist' in research on the user patterns of new media carried out by Van den Beemt[16];
• **networkers** – approximately one quarter of the pupils is of the 'networker' type (see Figure 10.3). Networkers are usually girls. They can be distinguished by their use of various kinds of social software – e.g. Hyves and MSN – for contact with friends;
• **gamers** – make use of all kinds of game applications such as pc-games, online games, console games and games on their mobile. In addition, gamers make use of the basic functions of internet, as do the other groups;
• **producers** – this group, the producers, is hardly found at primary schools (4%) but develops from the time they go to secondary school. These are pupils who use MSN, Hyves, YouTube, Flickr and MySpace to show others what is interesting or to obtain information on others.

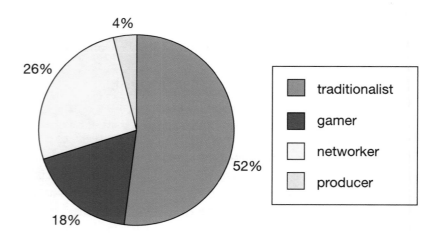

Figure 10.3 • Four types of users of new media in primary education.[17]

It is important for teachers to be able to recognise these four groups as it gives reference points for the pedagogical arrangement of the learning process. The four user groups have remarkably many similarities in their ways of processing information, i.e. learning styles. Traditionalists mainly consume text. Gamers are image oriented. Networkers mainly write text and producers make images, sound and text.

It is possible that there is stronger relation between media consumption on the one hand and learning styles on the other than has been charted to date. An understanding of the media consumption of pupils could give teachers the indication that in a class with many 'producers' it could be useful to use Web2.0 applications. This would, however, be of less use to 'traditionalists' or 'networkers'.

10.5 • YIELDS

Both teachers and managers are positive about the opportunities and yields of ICT. More than 80% of the school managers feel that ICT contributes to more appealing education. And ICT enables schools to offer pupils a richer learning environment.

The ICT yields can be expressed in different ways. Among other:
• lower purchasing costs for teaching materials;
• gain in teaching time;
• more appealing education, more motivated pupils;
• increasing the appeal of the teaching profession, reduction in teacher shortage;
• better learning results, reduction in learning disadvantages of disadvantaged pupils;
• other educational yields for pupils than learning results, e.g. command of skills for working together and independent learning;
• reduction of time spent at school and decrease in school dropout rate.

There is an increasing number of research results available with concrete proof that the above yields are realistic. Recent research has, for example, shown the following insights:
• ICT is very suited to offering teaching materials in different ways: visual, auditive and interactive. By offering teaching materials through several channels, pupils learn more and better;
• Particularly weak pupils often show a marked progress when they learn with the aid of ICT. A condition is, however, that well designed

programmes are used that are focused on the pupil's level, can steer the attention properly and can guide the pupil in steps through the material. If the digital teaching materials do not meet these conditions there is little or no effect;

• Pupils usually enjoy working with ICT and find it appealing. It is important, however, that ICT is alternated with other teaching types. A second crucial factor is that the attention of the pupil be directed at the learning tasks. It is therefore important not just to assess the ICT application on its appeal but also on the depth of learning and efficiency of tools used.

The above mentioned insights have been compiled in the framework of a research programme that has been carried out in the Netherlands by Kennisnet since 2007, called 'Kennis van Waarde Maken' (Creating Valued Knowledge). The result from this programme makes clear what does and what does not work in ICT in education. Brief video presentations of completed and ongoing research are available via onderzoek.kennisnet.nl.

MULTIMEDIA BOOKS

A striking example of the added value of new media for learning is the contribution made by multimedia books to the language development of pre-schoolers. Just like traditional picture books, multimedia books – also called 'living books' – tell a story. In contrast to 'normal books' this story is told on the computer screen thereby allowing for more interactive learning processes.

The events can also be illustrated with the use of sound fragments and music that enliven the story line. It is also possible to include exercises. Reading is then no longer static but dynamic.

Multimedia books help develop the language understanding among pre-schoolers (4 to 5 years of age), increase their vocabulary more quickly, teach the alphabet and greatly increase the reading skills.[18] It is sometimes thought that multimedia books can overload children's memory. Research, however, has shown the opposite. On the contrary, everything points to the fact that multimedia reading – words are presented as text, as sound and visually – helps children to understand more words.[19] Apparently the reference points and extra reference points to interpret words and understand stories have a leverage effect.

All children benefit from living books but, especially children with a language disadvantage are given an extra helping hand. Children with a language disadvantage often end up in a vicious circle when using traditional books. If your vocabulary is not good enough to be able to follow the story it becomes difficult to learn new words and you cannot really enjoy the story. The added value of multimedia is that it is an adequate tool to effectively break through this vicious circle, characteristic of children with a language disadvantage.

Generally, from an economic point of view, the yield of early intervention to combat a learning disadvantage or other disadvantages is much higher than the yield of intervention at a later age.[20] The example of the digital picture books has shown that particularly weak pupils for whom existing tools fail benefit the most from these ICT applications. The use of digital picture books for young children with a language disadvantage is not just a modern reproduction of a picture book. It is an innovation which can be used as intervention tool in education to bring about a substantial cost reduction in preventing or bridging language disadvantages.

The results of the research carried out over the past years on yields of ICT in education illustrate that ICT, when used properly, can provide a demonstrable contribution to more efficient, more effective and more appealing education. These appealing yields would in most cases not be attainable without ICT. At the same time, research results also show that not every ICT application necessarily leads to improved education yields.

In order to be able to make use of ICT in education on a larger scale, various barriers will have to be conquered, the most important of which is discussed below.

10.6 • MANAGEMENT VERSUS TEACHERS

Making effective use of new media for education purposes calls for more than just the availability of technology. The availability of new technology is necessary but by itself not the only condition needed to realise better education. In addition to material facilities that can be acquired by school management, the conquering of barriers with regard to the views and competencies of teachers also have to be dealt with. Although teachers and management share the ambition to make more use of new media in the future, in practice, a contrasting approach to attaining this goal has presented itself.

Management thinks that since the introduction of new media the majority of investments have been in infrastructural facilities. According to management, now is the time for a turnaround and the focus should now shift to a change in behaviour and views of the teachers. The teachers, however, have a different point of view as to what should be done. In order to be able to make more use of new media in the future, teachers' priorities go to adequate ICT facilities and useful digital learning material.

This contrast is not only found in primary education but also in secondary education and senior secondary vocational education.[21] It is one of the major barriers for further integration of new media in education. It is a situation in which the current role of new media in education is characterised as *oversold and underused*[22] by critics.

From the teachers' perspective the contrast can be seen in terms of internal and external reference points. External reference points indicate material preconditions, such as computers and digital teaching materials and internal reference points are knowledge, skills and views.[23] Individual teachers cannot break through the contrasts. This key role lies with management. In practice this means: developing a vision that can be supported by teachers and then closely following what works in practice and what does not. This kind of leadership, that offers teachers room to develop ownership and self-confidence in the pedagogical use of ICT, is lacking in their school according to most teachers.[24]

10.7 • FUTURE

Teachers and school management see ICT as a useful tool for improving the quality of the education system. The ambitions of schools is that pupils will be able to make more use of ICT for school work in the near future. There is now sufficient evidence to show that new media contributes in various ways to the quality of education: social, economic and didactic. The challenge for the future is to utilise the opportunities of ICT for improved education for more pupils. A first step in this direction can be made by making more use of facilities already present in the education system and less noncommittal use of ICT applications that have already shown their added value.

As mentioned, management and teachers differ in their views as to the approach to attain this goal. The solution to this contrast lies not so much in the focus on separate preconditions. In order to break through this contrast, a good balance within school organisations will need to be found between, on the one hand, the available material facilities, and on the other the views and skills of teachers. In other words: it concerns the alignment between the material preconditions that the school organisation can realise for the teachers and the professional competencies that teachers will to a great extent have to provide for themselves. It is the school management's responsibility to manage this alignment.

Following on from the analyses in other countries[25] it is becoming increasingly clear that the extent to which school management takes on ICT leadership is the most crucial factor for a greater yield from new media for better education in the future, at equal or even lower costs. An important lesson learnt from the past is that there is seldom a yield from new media if the acquisition of material facilities is the starting point for change. This approach can be called *technology push*. More likely to succeed is *educational pull,* in which the acquisition and use of new media originate from the pedagogical beliefs on the learning of children.[26]

Recent research underlines the importance of agreement between educational principles and ICT choice.[27] The challenge facing education is selective and targeted use of new media, attuned to educational content and pedagogical approaches. Meticulous tuning can result in the talents of more pupils in education being developed further.

NOTES

1 Maddux, C.D. & Johnson, L.D. (2009). Information technology in education: the need for a critical examination of popular assumptions. *Computers in the schools*, 26, pp.1-3.

2 Kanters, E., Vliet, H. van, Ringersma, D., Zwaan, M. & Kokkeler, B. 2009. *Web 2.0 als leermiddel. Een onderzoek naar het gebruik van nieuwe internettoepassingen door* jongeren (Web 2.0 as a learning tool. A study into the use of new internet applications by youngsters). (2009). Kennisnet Onderzoeksreeks (research series) no. 11. Zoetermeer, Stichting Kennisnet. Available via onderzoek.kennisnet.nl.

3 Bennet, S., Maton, K. & Kervin., L. (2008). The 'digital natives' debate: A critical review of the evidence. *British Journal of Educational Technology*. 39(5), pp.775-786.

4 Hawkridge, D. (1990). Computers in third world schools: The example of China. *British Journal of Educational Technology*, 21(1), 4-20;
Brummelhuis, A.C.A., ten (2006) Aansluiting onderwijs en digitale generatie (Connection of education with the digital generation). In J. de Haan & C., van 't Hof (eds.). *Yearbook ICT and society 2006*. The digital generation. Amsterdam: Boom (pp. 125-141).

5 AOI (1982). *Leren over informatietechnologie: noodzaak voor* iedereen (Learning about information technology: a must do for everybody) . Report of the Advisory Committee for Education and Information Technology. The Hague: Ministry of Education and Sciences.

6 Ministry of Education, Culture and Sciences (1993). *Decree on the organisation of vwo, havo, mavo, vbo. Info-series primary education, no.6*. Almere: Procesmanagement Basisvorming.

7 Raad voor Cultuur (2005). *Mediawijsheid. De ontwikkeling van nieuw burgerschap.* (Council for Culture. Media literacy. The development of new citizenship) The Hague: Raad voor Cultuur Available via www.cultuur.nl.

8 Ministry of Education, Science & Culture (2009). Policy response *advice on innovation of libraries 2009-2013*. Letter to the House of Commons. Available via http://www.minocw.nl/documenten/96993.pdf.

9 Annaniadou, K. & Claro, M. (2009). *21st century skills and competences for new millennium learners in OECD countries*. Paris: OECD.

10 Anderson, R. (2008). *Implications of the information and knowledge society for education*. In: J. Voogt & G. Knezek (eds.). International handbook of information technology in primary and secondary education. New York: Springer;
European Communities (2007). *Key competences for lifelong learning*. Luxembourg: Office for official publications of the European Communities. Available via http://ec.europa.eu/dgs/education_culture/publ/pdf/ll-learning/keycomp_en.pdf;
OECD (2006). *The new millenium learners: challenging our views on ICT and learning*. Paris: OECD-Ceri. Available via http://www.oecd.org/data-oecd/1/1/38358359.pdf.

11 Kennisnet (2009). *Vier in Balans Monitor (Four in Balance Monitor) 2009*. Zoetermeer: Kennisnet. Available via onderzoek.kennisnet.nl.

12 Kennisnet (2009). See note 11.

13 Beerepoot, R, Leenaerts, M., Meij, E. van der., Mokkink, I. & Sprenger, J. (2009). *ICT-gebruik in primair onderwijs* (the use of ICT in primary education). Utrecht: Berenschot.

14 Rooij, A.J. van (2008). *Monitor Internet en jongeren* (Monitor Internet and youngsters). Rotterdam: IVO. Available via onderzoek.kennisnet.nl.

15 Intomart GfK (2009). *Customer satisfaction survey Kennisnet*. Hilversum: Intomart GfK.

16 Beemt, A., van den. (2009). *Jongeren en interactieve media (Youngsters and interactive media)*. Kennisnet Research series no. 17. Zoetermeer: Kennisnet.

17 Beemt, A., van den (2009). See note 16.

18 Bus, A.G. (2009). *Wat weten we over ict en taalontwikkeling van jonge kinderen* (What do we know about ICT and language development of young children). Kennisnet research series no. 15. Zoetermeer: Kennisnet. Available via onderzoek. kennisnet.nl.

19 Verhallen, M.J.A.J., Bus, A.G., & Jong, M.T. de (2004). *Elektronische boeken in de vroegschoolse educatie* (Electronic books in pre-school education). Available via www.lezen.nl.

20 Heckman, J.J. (2006). Skill formation and the economics of investing in disadvantaged children. *Science*, 312:5782, pp. 1900-1902. Available via www.brainwave. org.nz/wp-content/uploads/2007/03/valuing_prevention.pdf.

21 Kennisnet (2009). See note 11.

22 Cuban, L. (2001). *Oversold and underused: computers in the classroom*. Cambridge, MA: Harvard University Press;
Hixon, E. & Buckenmeyer, J. (2009). Revisiting technology integration in schools: implications for professional development. *Computers in the Schools*, 26:2, pp.130-146.

23 Lu, R. & Overbaugh, R.C. (2009). School environment and technology implementation in K-12 classrooms. *Computers in Schools*, 26: pp.89-106.

24 Kennisnet (2009). See note 11.

25 Byron, E. & Bingham, M. (2001). *Factors influencing the effective use of technology for teaching and learning*. Eric document reproduction service No. 471140;
Lu & Overbaugh (2009). See note 23.

26 Brummelhuis, A.C.A. & Kuiper, E. (2008). Driving forces for ict in learning. In J. Voogt & G. Knezek (red). *International handbook of information technology in primary and secondary education*. New York: Springer.

27 Mishra, P. & Koehler, M.J. (2006). Technological Pedagogical Content Knowledge: A new framework for teacher knowledge. Teachers College Record. 108, pp.1017-1054;
Tondeur, J. (2007). Development and validation of a model of ICT integration in primary education. Gent: PhD thesis, Universiteit Gent.

"We prefer all using the computer together"

Pauline Maas is a multimedia teacher. She teaches children, among other, how they can make their own computer games. She is also the mother of four adolescents. Her youngest, Vivianne (12), thinks it's cool that she has such a clever ICT mum. This doesn't, however, mean she can spend the whole day on the computer.

Pauline: "Vivianne prefers to draw and make clothes, she doesn't spend more than an hour per day online. So I don't need any strict rules for her, but for the others this is sometimes necessary. Her brother is a fanatical player of games. I only allow him to do that for two hours per day. I work from home and have three computers. After school we all sit in my office together. I like to know what they are doing and it's also fun. If they have questions on their homework or the computer they can ask and we show each other crazy films or nice pictures. We have a big house but we prefer to all sit crammed up together in my office, ha ha."

"I made a Hyves page together with Vivianne and explained to her what she should and shouldn't do. Passing on your password, for example, or adding strangers or putting certain photos online. I know children who have more than 200 friends on Hyves, I think this is risky. People they don't know can see everything they are doing."

"I teach children at schools how they can make animations and games, or how you can operate robots with the computer."

"I also let Vivianne try out things sometimes. She enjoys that but my other daughter would have preferred a 'normal' mother. She talks about Twitter and I already know everything or I'm not allowed to have anything to do with her Hyves, ha ha."

Vivianne: "I actually think it's great and relaxed that my mum knows so much. Everyone was talking about Hyves and I could just ask her what it is. My classmates' mums really don't know things like that. I'm very proud of her."

"We have three computers. One is my mum's, I share one with my sister and my brother and other sister also share one. I'm the youngest and the only one without homework so I'm often kicked out. And then my sister, who's supposed to be studying, goes and watches a missed TV programme! Oh, it doesn't really bother me. Some children have their own computer and spend all day on it, I can't understand that parents allow this."

"I love playing games, such as Harry Potter, or The Sims. Sometimes I buy CD-ROMS and I also play on the Wii. But after a while I've had enough of the computer. Friends usually tell me: 'you should do this!' Then I install it or create an account and after a month I'm fed up with it. I'm not very active on Hyves, either. I think other people think it's cool to have so many friends. They proudly call: 'Hey yo, I've already got 245 friends!' When I tell them that I've got 36 they look at me strangely. I used to have the Dutch actress Nicolette van Dam as friend until my mum said: 'you don't know her.' So I removed her. And besides, how do you know whether those famous people on Hyves are really who they say they are?"

- www.hyves.nl
- www.spelletjes.nl
- www.spelen.nl
- www.4pip.nl

11

Parental Mediation

Peter Nikken
Justine Pardoen

Parents are busy with all sorts of things on a daily basis: cooking, shopping, taking and picking up children from school, laundry, tidying up and keeping an eye on what their children do on the computer or the internet. With all the television programmes, game consoles, mobile phones and computers on offer, parental mediation of children's media use has become a serious part of the traditional upbringing.

In this chapter we will discuss what research has taught us about parental mediation. We will discuss how media guidance by parents compares to the general upbringing and how effective mediation can be. Finally, we deal with the needs among parents for support in mediation and what tools are currently available to support the upbringing.

11.1 • WHAT IS PARENTAL MEDIATION?

Parental mediation is the part of the parental upbringing aimed at enabling children to consciously and selectively manage the media on offer and seeing to it that they can take a critical look at the content on offer and are able to assess its value.[1] In daily practice this involves children's gaming, using the internet, reading and watching TV. Parents do not always consciously guide their children but what they do – or do not do – always has an influence on the growing up of children. The decision, for example, on whether to buy a *Nintendo Wii Fit* or not, the amount of time parents themselves spend on the internet or reading and the enthusiasm with which they text their vote for their favorite contestant in their favorite contestant in programmes such as *Het beste idee van Nederland* (Holland's Best Innovation), are all signals for their children on how to deal with media.

Parents try to fill in their parental role to the best of their abilities, knowing full well that they are responsible for their child.[2] All parental behaviour and statements are inspired by the (implicit or explicit) expectations of parents of their family and their children. There are, therefore, differences in the way parents guide their children in dealing with media. Parents will, for example, make other choices on the use of games and the computer for young children than they do for older children. Parental mediation for girls can also be different than for boys. Parents are usually more inclined to disapprove of shooting games for girls, while boys have more freedom in this.[3]
All is dependent on the circumstances and on what parents consider to be good and valuable for their children in those situations.

But bringing-up is not purely a matter of setting an example with regard to standards and values and stimulating or intervening in the development of children. It is also a learning process for the parents themselves.[4] Children acquire autonomy as they grow older, particularly through 'disobedience dialogues' with parents or other carers. By constantly trying out how far they can go, children push back boundaries. Bringing up children is, therefore, a continuous process of endless repetition in constantly different situations, for both the parents and the children. Parents constantly have to reconsider what they should do in a specific situation with or for their children. They then discover whether their handling of the situation was sufficiently pedagogic. Growing up and bringing up thus takes place through relations that family members have with each other and that are constantly in motion.[5] Children are not passive receivers of parental care and socialisation but are themselves (moral and social) participants of the family practice.

Also when dealing with the media, parents constantly sound out their child, set boundaries, do or do not do something together with their child and discover afterwards if their behaviour was satisfactory. This does not always take place rationally or consciously. Upbringing often takes place on the spot. If parents are fed up with their children having spent too much time on the computer, they can suddenly become angry. In other cases, there is more peace and quiet and they can regulate their emotions better. It is also possible that parents let the situation take its course for a while. Other matters require attention in which case it is more convenient for the parent to give the child his way. That does not, however, mean that the parent is then happy with the situation. On the contrary, the parent may feel very guilty and feel that they have fallen short in their parental mediation. Contemplating on these kinds of situations and thinking what should be done next time when a situation occurs, ensures that the upbringing doesn't fail.[6] This gives parents confidence in their parental role. It is also vital that parents divide the rearing tasks "well". If they do not, and also insufficiently reflect on their media guidance, there is a possibility that children show high-risk media behaviour.

Although the upbringing of children is also for parents a process of trial and error, the upbringing process does have five essential acts:
• keep an eye on the child;
• offer security;
• offer physical and psychological care;
• guide children by expressing expectations;
  set boundaries.[7]

All parents do these things during the period in which they are responsible for the upbringing, although content wise this changes somewhat as the children grow older. Different successive phases can be distinguished regarding the way in which parents and children interact with each other as the children grow up: showing, discovering together, coaching children and, finally, letting go.

The way in which parents guide the media use of their children corresponds, in part, with the above-mentioned five upbringing tasks. According to game and television research, parental mediation consists of three styles:[8]
• *restrictive mediation:* regulating what children do with media, e.g. forbidding certain programmes or games, making agreements on time spent and when children are allowed to do something. Parents also set boundaries for media;
• *active mediation:* exchanging opinions, comments and information about what children stumble across in the media, e.g. giving reasons why certain games are 'good' or 'bad', explaining subjects in the me-

dia that children cannot yet understand properly or applauding nice and suitable media expressions. This supervision is also called evaluative, instructive or informative mediation and corresponds to guiding children by expressing expectations;
- *social coviewing or coplaying:* consciously watching or gaming together, parents and children enjoy themselves, get the creeps or sympathise together. The important thing is exchanging emotions during the use of media together, although a discussion also takes place on what is fun or interesting. Some researchers feel that collective media experience is not really bringing up because it takes place unintentionally.[9]
- From the child's perspective, however, it can most certainly be effective. Cuddling up to a parent when a scene in a fairy tale becomes too scary can help young children curb their fears of media productions.

SUPERVISION ON THE INTERNET

Research by Stichting Mijn Kind Online (My Child Online Foundation) and Motivaction among 621 Dutch parents of 6- to 12-year-olds[10] has shown that the three styles of parental mediation are also found in the internet use of 6- to 12-year-olds:
- parents join their children at the computer and exchange experiences. Parents ask their children what they experience on the internet and what they like;
- parents set boundaries. They make agreements on how long and when children may use the internet. In addition, they determine which sites or games may be visited or which music they can download;
- parents guide their children by telling them what is right and wrong on the internet. They indicate what their children should do when keeping in touch with others on Hyves or MSN. Parents give explanations and help their children to think about their own safety.

Besides these three types of mediation, parents of children up to the age of 12 also use a fourth type of supervision: parents give their children freedom on the computer but make sure that they know what their child is doing on the computer, from a distance. This type of supervision, keeping an eye on things, is most used by parents (see Figure 11.1). It appears that more than half to almost three-quarters of the parents with children aged between 6 and 12 regularly or often stay in the vicinity to be able to help should that be necessary or keep an eye on what is happening on the computer in the living room. Apparently, the general task of upbringing, watching what a child does, parents also take very seriously when it comes to media.

After 'keeping an eye', going on the internet together and exchanging experiences is the second most frequently applied type of mediation for children aged between 6 and 12. Asking children about their experiences on the internet and surfing together is done often or very regularly by approximately 1 out of every 3 parents. Next, actively explaining what safe behaviour is in contact with others is also something a fair number of the parents with children up to the age of 12 do. Finally, setting rules for internet use comes last, on average. Slightly fewer than half of the parents say they seldom or never forbid specific surfing behaviour on the internet. Rules on the amount of time spent and when and which games are ok, on the other hand, are still set fairly often.

Parents also try to ensure the safety of their children on the internet using other measures. More than 30% to almost 60% of the parents have standard services such as a firewall, a virus scanner, spam filters, popup killers and filters for unsuitable websites. Specific services for children are found less often: no more than 9% of the parents has a monitor for the internet behaviour of their children and 6% operates a white list, a time clock or a chat room filter.

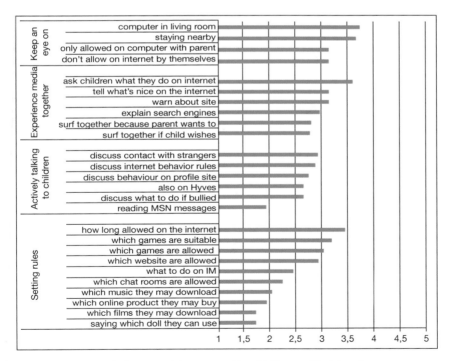

Figure 11.1 • Degree to which parents of 6- to 12-year-olds guide the use of the internet

(1 = never; 5 = very often). Source: Nikken (2009); N = 621.

For *stand alone* games on a game console or a PC, parents often use restrictive mediation. 63% of the parents regularly check which PEGI age classification a game has, 56% keeps an eye on the games their children play and 46% get their children to ask permission before being allowed to play games.[11] A quarter to more than a third of the parents ensure active guidance. For example: regularly discussing the content, explaining, pointing out the good and bad things of a game and drawing attention to 'fakes'. Experiencing games together occurs least often, relatively speaking. Merely one out of every eight parents consciously play a game offline together with their children regularly.

11.2 • MUCH OR LITTLE PARENTAL MEDIATION

Some parents are more active in their guidance of the media use of their children than others. Mediation is generally taken care of more by mothers than by fathers. This has been shown for watching TV and gaming[12] but also for use of the internet.[13] Particularly making agreements on what is and is not allowed and talking about 'good' and 'bad' in the media is something mothers do more often than fathers. This is possibly because mothers traditionally fulfil the function of gatekeeper in the family. Mothers are at home more often than fathers and in the division of tasks they are more actively involved in the guidance of their children.

Highly educated parents also interfere with the TV and the internet behaviour of their children more often than less well educated parents. They indicate more often what they are allowed, keep an eye on this more often, and talk to their children about media more often. For offline games this is, however, different. In this case, the less well-educated parents are the ones supervising more.[14] It is possible that these parents are more active in gaming as game consoles are more common in less well educated families. A requirement in being able to give children guidance with media is, after all, that the carers know what they are talking about.[15] Parents who are skilled internet users are more inclined to talk to their children about how they can behave safely on the internet than parents with fewer internet skills.[16]

VIEWS ON MEDIA EFFECTS

Those parents with more personal experience with media than other parents often have more well considered expectations of media. Their personal media experience allows them to more easily assess whether their children can cope with certain media expressions and what the effects of the media can be on their children. According to parents, there are roughly two types of effects that can be attributed to the media:[17]

- On the one hand, parents expect positive media influences on their children. This mainly includes cognitive effects: gaining knowledge on nature, language development, stimulation of concentration and thinking ability as well as improvement of hand-eye coordination;
- On the other hand, parents are worried about possible negative effects on their children, such as violent behaviour, rude language, getting a wrong world-view, bullying via the internet or having too little time left for other activities.

The view of parents on the effects of media is usually nuanced. At the same time parents recognise the positive and negative influences. There are few parents who only see the positive or only see the negative effects. On average, parents do, however, often think the media are more likely to positively contribute to the development of children rather than it being harmful to their growing up.

In general, parents worried about the negative effects are more likely to set boundaries for the media use of their children.[18] If parents think media can be harmful for their children, they are not allowed to watch as much television, there is more control on offline games and more rules are set at home on what children may and may not do on the internet. Parents tend to stay closer to their child when on the internet, 'just in case'. Besides, parents worried about the harmful effects of media talk to their children more often about what they experience in the media. Parents with a positive view of the media are also slightly more inclined to talk with their children about it. Above all, however, they feel that it is important to watch, play games and surf together with their children.

If parents are worried about their children's gaming, it usually has to do with the amount of time it costs.[19] More than 11% of the parents with children aged between 6 and 9 think their children spend too much time gaming and of the 10- to 12-year-olds the concerned group grows to almost 20%. Parents of young children see that the gaming harms their social contact with other children while older children lag behind in their homework.

Parents make a distinction between boys and girls in their media guidance, and also between older and younger children. Previous research on parental mediation has shown that parents set more boundaries for the gaming and TV-watching of their daughters than of their sons.[20] For the use of the internet, parents hardly differentiate between sons and daughters.[21] Parents set rules on what may and may not be done on the internet equally often for boys and girls aged between 6 and 12. They also stay in the vicinity to keep an eye on everything just as often, surf together equally often and talk to their sons and daughters equally frequently about safe internet behaviour. This probably changes as children grow older.[22] In addition, there are also specific internet behaviours for which parents do make a distinction between sons and daughters: girls in grade 6 have many more rules set with regard to what they can do on Hyves than boys.[23]

Parents also supervise younger children in another way than older children. For younger children parents keep an eye on the gaming of their children, talk about it and also often game together.[24] Regulating the watching of TV and talking about programmes together are also things parents do more often with younger children than for children ready to go to secondary school.[25] For the use of the internet, however, parents seem to grow along with the children, at least for as long as the children go to primary school.[26] Parents talk about what children can do to behave in a safe manner on the internet more often with children in the upper level of primary school than with younger children. In addition, parents also set more boundaries for the internet behaviour of the older children but on the other hand, they then stay in the vicinity to keep an eye on things less often. More intensive media guidance for older children coincides with them surfing more directly on the internet for content that they can look at, listen to and download; and being more busy maintaining social contacts via the web. Parents apparently think it important to steer this surfing behaviour, in particular, in the right direction.

11.3 • WHAT WORKS?

Over the past years, research has been carried out in both the Netherlands and abroad on the possible outcomes of the media guidance by parents. Research on the effectiveness of mediation, to date however, gives a limited picture, partly because the studies have been carried out among families with children in different age groups and for different media. Besides, research on parental mediation has up to now only been carried out among 'normal' children. Nothing is known about the mediation of, for example, children with an attention disorder (ADD or ADHD) or a social interaction disorder (ASD).

There was also hardly any research that shows what the effects were of the enormous expansion of the media on offer over the past decades on parental mediation and on the functioning of families in general. It is possible that the communication types among children and between children and parents is changing because of the new media types and because children and their parents now spend less time together than when they only had television. Nor do we know how parents divide their media guidance tasks if they are divorced or if children are confronted with the differing views of biological parents and stepparents or if there are other problems that can influence the upbringing.

Taking all studies on parental mediation into consideration, however, it is plausible that the intervention by parents matters, although mediation can, in some cases, also be counter productive.[27]

RULES

Making agreements with children on things that they may and may not do on the computer generally proves to be effective in combating negative media effect, at least for children older than 12 years of age.[28] But then it is important that parents know what they want to protect their children from. Putting filters on the computer and assuming that the internet then automatically becomes safer is not enough.[29]

Whether parental internet rules are also effective for children under the age of 12 has scarcely been studied. It is known that Korean children who have been set rules as to the amount of time they are allowed to spend on the internet has *not* resulted in actually spending less time on the internet.[30] On the other hand, studies carried out on children who had their TV-watching or offline or online gaming restricted or where parents controlled the media use, have shown that these children less often played computer games for which they were too young, were less violent and more prosocial in their behaviour, reacted less terrified to violence in the news and were less likely to agree with media violence.[31] In addition, children performed better at school if their parents see to a healthy balance between the use of new media and other behaviours.[32]

Relatively speaking, most research has been carried out on the active interference of parents in the media behaviour of their children. Most studies show that active mediation in which parents talk, give information or explanations or give their own opinions to their children, is effective. Children who are actively guided are more critical of and more involved with news items, they take fictional violence less seriously, they are more aware of violence in society and they react less terrified and worried to serious news items.[33] Furthermore, Dutch research has shown that active guidance can also result in children becoming more critical of products in advertising and asking for these products less often.[34] For children in their puberty it has also been shown that if their parents regularly address them regarding their internet behaviour they show less compulsive internet behaviour six months later.[35] Media use can in this way be kept within limits. That active guidance of internet behaviour is also effective for younger children, is plausible but very little research has been carried out yet on this age group. For Korean children who are actively shown suitable websites by their parents, it has been shown that they do indeed surf to educational websites more often.[36]

Talking to children about media is not per definition an effective way of upbringing. It depends on *how* parents communicate with their children on media. In fact, there are indications to show that parental interference can also be counter productive. Adolescents who get many comments and criticism from their parents on the duration of their gaming behaviour and on the content of the games they play, while at the same time their parents set very few rules, play games with a PEGI classification that is too high most often.[37] A possible explanation for this phenomenon could be that parents have watched the gaming behaviour of their children get out of control with concern and now finally dare to say something. On the other hand, it is also possible that the criticism from their parents is what makes them play these games, the so-called 'forbidden fruit' effect. Particularly the latter is plausible as children quickly see through their parents, if parental comments are mainly negative, based on unreasonable fears (at least in the child's view unreasonable) or on lack of knowledge.

The results of active mediation style are strongly related to the quality of the discussions. Discussions started positively and a warm and open relationship between parent and child, are generally conducive for the healthy development of the child. Particularly during disobedience dialogues parents can air their expectations and guide their children to deal with media in a critical and conscious way. Children with involved parents with whom they talk about media and their experiences with it do not just have a more positive attitude towards the media but also appreciate being able to talk to their parents about it.[38]

EXPERIENCING MEDIA TOGETHER

Watching TV together, gaming or going on the internet together can have positive results especially if parents keep the positive effects of the media for their children in mind. As far as the research on experiencing media together permits, it is also possible that watching together or gaming or going on the internet together could work counter productively.

Children who often play games together with their parents, are more inclined to show aggressive behaviour, are more likely to approve of media violence, play games for which they are too young more often and more often think that their parents approve of the violence in games.[39] These results, however, always were found in surveys. It is therefore questionable whether these behaviours and attitudes among children are the result of regularly watching or gaming together or that parents decide to watch or play with their children more often because of their behaviour.

11.4 • WHAT DO PARENTS WANT?

In 2005, the Dutch Raad voor Cultuur (Council for Culture) recommended that children and their carers should learn how the influences of the media work, how they can differentiate between sense and nonsense and how they themselves can use the media for their own purposes.[40] Carers and children have to be 'media-wise' and preferably as young as possible. In 2005, the Commissie Jeugd, Geweld en Media (Youth, Violence and Media Committee; JGM) also advised the government on policies with regard to children and media. This committee reported, among other, that the present-day media landscape puts a lot of pressure on the upbringing task of parents.[41]

Media are everywhere and often already outdated by the time parents have managed to get to grips with it. In addition, children see examples in the media that can be opposed to the upbringing ideals of their parents. According to the Committee JGM, it is more difficult for parents nowadays to fulfil the traditional gatekeepers function. Media guidance has become more complex. There are so many media which children can consume increasingly often, whether on their own, at friends or at the after school care. Besides, no parent wants his child to miss out on happiness (i.e. entertainment) and educational opportunities. It is therefore more important than ever that parents guide their children on media and help their children develop a conscious and critical attitude towards media. According to the Committee JGM, the need of parents for information on children and media is great.

Parents of children aged between 6 and 12 have distinct views on the role that they play as mediators and what is needed. Slightly more than half of the parents feel it is primarily their task to guide their children on the internet while almost all other parents feel that parents and school should be jointly responsible for media literacy.[42] It is for parents also a fait accompli that the media, in particular the internet, is part of a child's life (see Figure 11.2). Children have to learn to deal with this and the education system also has to pay attention to media literacy.

In addition, a large number of parents also thinks that primary education should spend more time on media literacy than is now the case, so that children can safely use the internet. According to parents, schools use media and ICT more as tools in education rather than for education on media. Only one in every four parents of children aged between 6 and 12 have noticed that their child has learnt how to use internet safely

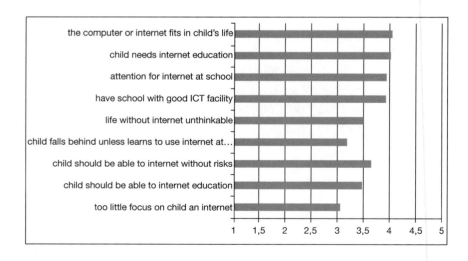

Figure 11.2 · What parents with children aged between 6 and 12 think about school and the internet (1 = totally disagree; 5 = agree totally). Source: Nikken (2009); N = 621.

and in a protected way. These results can also probably be extended to child care. Children can, after all, also play games, surf and watch films and television programmes there.

Parents feel that parents' evenings dedicated to the internet and other media at primary schools are a good way to learn about media literacy. Parents feel that parents' evenings and contact with other parents (e.g. at the school playground) are better sources of information and knowledge than national campaigns, books or research reports. Parents feel the need for low-threshold contact where they can exchange their own experiences and knowledge with other parents and professional carers.

Information that should be available to parents regarding support on the internet behaviour of their children should include three things: [43]
- Firstly, parents want to know which sites and games are suitable for their children. With the enormous number of websites, television programmes, DVDs and games on offer, parents want a guide to help them get a clear picture;
- Secondly, parents want recommendations on how they can best support their children in maintaining their contacts via the internet. Parents want to know what is sensible to do if their child has contact with strangers, what they should do if personal details are asked for and what information children should not give on their Hyves' profile sites;
- Finally, parents, although to a lesser extent, are interested in information on responsible downloading and the use of games, films and music.

11.5 • WHAT IS AVAILABLE FOR PARENTS?

For both parents and professionals concerned with educational support it is not difficult to find information on the internet on children and media. Both commercial and non-commercial organisations offer websites, brochures, factsheets, research reports and even complete parents' evenings on children and media.

It is more difficult for parents to determine whether the available information is also useful and usable for their own media guidance situation. How reliable is the information? Institutes that provide information on children and media and give advice to parents have no official quality mark. There is also no trade union or trade association for advisors on media literacy. Parents looking for relevant information will really have to search and hope that the information they find meets their own upbringing ideals.

Although parents will mainly have to fall back on their own intuition and common sense, it does not mean that they are totally on their own. Besides the odd initiatives of many organisations and institutes, the Dutch government has also started various developments to support parents over the past years: Expertise Centre Media Literacy, Digivaardig & Digibewust (Digi skilled & Digi aware), Kijkwijzer / PEGI, and MediaSmarties.

EXPERTISE CENTRE MEDIA LITERACY

Since 2008, an Expertise Centre Media Literacy has been established to help children and adolescents, their parents and teachers in learning to deal with the large amount of media expressions. One of the main tasks of the expertise centre is bundling and providing an overview of all information sources available on media and children, education, upbringing, youth work cultural participation, etc., and identify what gaps there are.

Realising a public function, with counters open to parents, still has to materialise. They are currently thinking of the many national libraries, the youth and family centres and the Nederlands Instituut voor Beeld & Geluid (Netherlands Institute for Sound and Vision) in Hilversum. In time, parents should be able to get ready-made answers to questions on media literacy at such locations.

DIGIVAARDIG & DIGIBEWUST

The government, in collaboration with trade and industry, is also active in the programme 'Digivaardig & Digibewust' (Digi skilled & Digi aware). First of all, the government wants everyone to acquire sufficient digital skills to be able to take part in the medialised society. Secondly, they want to stimulate the responsible and safe use of digital means. The programme is specifically aimed at the digitally illiterate including employees in small and medium sized companies and senior citizens, but also adolescents and their carers, such as parents and teachers. The programme 'Digivaardig & Digibewust' is available via the website Mijndigitalewereld.nl.

after withdrawing the film inspection, a kind of co-regulation was created, which made the media industry together with the government responsible for supporting parents. All films, DVDs, TV programmes and games are age rated. Furthermore, the use of one or more content symbols indicates why a product can be harmful for children up to a certain age. The pictograms clearly show parents whether a film, TV programme, DVD or game contains possibly harmful elements for their child. For films, DVDs and TV programmes this classification system is called the Kijkwijzer, for games this is 'PEGI' (Pan-European Game Information). In the Netherlands both classification systems are the responsibility of the NICAM, the Netherlands Institute for the Classification of Audiovisual Media.

Parents are sometimes surprised by the age rating of a certain game, film or TV programme. If parents wish to object they can inform the complaints committee of the PEGI or Kijkwijzer. Everyone who wishes to submit a complaint can do so via Kijkwijzer.nl or Pegi.info. An independent council will monitor whether the age and content rating was appropriate. If this is not the case, the broadcaster or distributor may get a fine. In addition, the rating has to be corrected as soon as possible. Parents thus have great influence on the workings of Kijkwijzer and PEGI.

MEDIASMARTIES

A final government initiative to ease media literacy for parents and professional carers was started in September 2009: MediaSmarties, a project to systematically give information on the suitability (rather than harmfulness) of media products for children. Parents do not just want to know for which products their children are still too young, they also want to know which DVDs, games, websites or TV programmes are suitable for their children.

The MediaSmarties system will make this information available via the internet. Professional editors will make objective reviews of programmes, films, games and websites taking into account the development and media use of children of different ages and give parents made-to-measure recommendations. In addition, parents can also react to the advice given thereby helping other parents to also choose suitable media products.

11.6 • CONCLUSION

Since the arrival of radio and television, media guidance has become part of the traditional upbringing practice. But in the current, strongly medialised society bringing up children has not only become more complex than ever before, parental mediation of children's media use is also more necessary than ever. Children can choose from an enormous range of media productions, whether or not suitable for them. In addition, children also have many more and technically more advanced media devices at their disposal than in the past, on which they spend much more time per day and which they keep more to themselves and out of their parents' sight.

These developments call for more insight into what good media guidance involves. Apart from professional carers, parents also have the responsibility to help children become media-wise. Within the research on parental mediation, consensus has to a large extent been reached on three styles of upbringing that parents can use to educate their children on media: restrictive mediation, active mediation and social co-use; i.e. intentionally viewing, surfing or gaming together.

At the same time there are also many blanks and little insight into how mediation in the family takes place exactly and which acts by parents are really effective. Particularly what parents do in specific situations, e.g. for children with a disorder, in divorced or reconstituted families or if other problems put the upbringing under pressure, is still unclear.

Further research is also necessary on how parents deal with the multimedial family life. How do they divide their attention on how their children deal with television, games, cinema films and the internet? Do they set requirements for each of these media or do they overlap?

Finally, a better understanding of the parents' needs with regard to support in their parenthood is needed. Which information sources on children and media do they use, are these sources effective and do they sufficiently fit in with their upbringing wishes? It is important to know what role tools such as Kijkwijzer and PEGI play in the daily practice of parental mediation.

NOTES

1   Nikken, P. (2002). *Kind en Media: Weet wat ze zien* (Child and Media: Be aware of what they see). Amsterdam: Boom.

2   Van der Pas, A. (2003). A serious case of neglect. The parental experience of child rearing. Delft: Eburon Academic Publishers.

3   Nikken, P. (2007). *Mediageweld en kinderen* (Media violence and children). Amsterdam: SWPbooks.

4   Van der Pas (2003). See note 2.

5   Hoek, M. (2008). Ontheemd ouderschap; Betekenissen van zorg en verant-woordelijkheid in beleidsteksten opvoedingsondersteuning 1979-2002 (Uprooted parenthood; The meaning of care and responsibility in policy texts on support of upbringing 1979-2002). PhD thesis University of Utrecht.

6   Hoek (2008). See note 5.

7   Hoek (2008). See note 5.

8   Valkenburg, P., Krcmar, M., Peeters, A. & Marseille, N. (1999). Developing a scale to assess three styles of television mediation: Instructive mediation, restrictive mediation, and social coviewing. *Journal of Broadcasting & Electronic Media, 43*, pp. 52-66.
    Nikken, P. & Jansz, J. (2006). Parental mediation of children's videogame playing: A comparison of the reports by parents and children. *Learning, Media and Technology, 31*, pp.181-202.
    Vittrup, B. What US parents don't know about their children's television use: Discrepancies between parents' and children's reports. *Journal of Children and Media, 3*, pp.51-67.

9   Valkenburg et al. (1999). See note 8.

10   Nikken, P. (2009). *Ouders over internet en hun kind* (Parents on the internet and their child). Den Haag: Stichting Mijn Kind Online (My Child Online Foundation)/ Motivaction.

11   Nikken, P. (2002). *Kind en Media: Weet wat ze zien* (Child and Media: Be aware of what they see). Amsterdam: Boom.
    Nikken & Jansz (2006). See note 8.

12   Nikken & Jansz (2006). See note 8.

13   Nikken (2009). See note 10.

14   Nikken & Jansz (2006). See note 8.

15   Austin, E. (1993). Exploring the effects of active parental mediation of television content. *Journal of Broadcasting & Electronic Media, 37*, pp. 147-158.

16   Nikken (2009). See note 10.

17   Mendoza, K. (2009). Surveying parental mediation: Connections, challenges, and questions for media literacy. *Journal of Media Literacy Education, 1*, pp.28-41.

18   Nikken (2009). See note 10.
    Valkenburg et al (1999). See note 8.
    Nikken, P., Jansz, J. & Schouwstra, S. (2007). Parents' interest in videogame ratings and content descriptors in relation to game mediation. *European Journal of Communication, 3*, pp. 315-336.
    Van der Voort, T., Nikken, P. & Van Lil, J. (1992). Determinants of parental guidance of children's television viewing: A Dutch replication study. *Journal of Broadcasting & Electronic Media, 36*, pp. 61-74.

19 Nikken, P. (2003). Ouderlijke zorgen over het 'gamen' van hun kinderen. (Parental concerns about their children's gaming). *Pedagogiek, 23*, 303-317.

20 Mendoza (2009). See note 17
   Nikken & Jansz (2006). See note 8.

21 Nikken (2009). See note 10.

22 Livingstone, S. & Helsper, E. (2008). Parental mediation and children's Internet use. *Journal of Broadcasting & Electronic Media, 52*, pp.581-599.

23 Duimel, M. (2009). *Krabbels & Respect plz? Hyves en kinderen* (Scribbles and Respect plz ? Hyves and children). Den Haag: Stichting Mijn Kind Online (My Child Online Foundation).

24 Nikken, P. (2003).See note 19.

25 Van der Voort et al. (1992). See note 18.

26 Nikken (2009). See note 10.

27 Mendoza (2009). See note 17.

28 Livingstone & Helsper (2008). See note 22.
   Duimel, M. & de Haan, J. (2007). *Nieuwe links in het gezin* (New links in the family). The Hague: Sociaal en Cultureel Planbureau (Netherlands Institute for Social Research).
   Van den Eijnden, R., Spijkerman, R., Vermulst, A., Van Rooy, T. & Engels, R. (2009). Compulsive internet use among adolescents: Bidirectional parent-child relationships. *Journal of Abnormal Child Psychology*. http://www.springerlink.com/content/98428w4524480305/fulltext.pdf

29 Mesch, G. (2009). Parental mediation, online activities, and cyberbullying. *CyberPsychology & Behavior, 12*, pp.387-393.

30 Sook-Jung, L. & Young-Gil, C. (2007). Children's internet use in a family context: Influence on family relationships and parental mediation. *CyberPsychology & Behavior, 10*, pp.640-644.

31 Nikken (2007). See note 3.

32 Gentile, D. & Walsh, D. (2002). A normative study of family media habits. *Applied Developmental Psychology, 23*, pp.157-178.

33 Nikken (2007). See note 3.

34 Buijzen, M. & Valkenburg, P. (2005). Parental mediation of undesired advertising effects. *Journal of Broadcasting & Electronic Media, 49*, pp.153-165.

35 Van den Eijnden et al (2009). See note 28.

36 Sook-Jung & Young-Gil (2007). See note 30.

37 Nikken, P. & Jansz, J. (2007). Playing restricted videogames: Relations with game ratings and parental mediation. *Journal of Children and Media, 1*, pp.227-243.

38 Mendoza (2009). See note 17.

39 Nikken (2007). See note 3.

40 Mediawijsheid: De ontwikkeling van nieuw burgerschap (Being media-wise: The development of new citizenship). The Hague: Raad voor Cultuur (Dutch Council for Culture).

41 Valkenburg, P. (2005). Schadelijke media, weerbare jeugd: Een beleidsvisie (Harmful media, resilient youth: A vision on policy). In A. van der Stoel, N. van Eijk, D. Hoogland, E. van Noorduyn & M. Wermuth. *Wijzer kijken: Schadelijkheid, geschiktheid en kennisbevordering bij het gebruik van audiovisuele producten door jeugdigen (Wiser in watching: Harmfulness, suitability and knowledge advancement*

with regard to the use of audiovisual products by youngsters). Advies Commissie Jeugd, Geweld en Media (Advice by Youth, Violence and Media Committee). pp. 32-49. The Hague: OCW.

42   Nikken (2009). See note 10.

43   Nikken (2009). See note 10.

12

Trends, conclusions and recommendations

Jos de Haan
Remco Pijpers

Since the start of the advance of the internet, and certainly after the arrival of broadband internet, carers and teachers have asked themselves the question what children should learn to deal with new media in a safe and effective way. Which skills are necessary to make media-literate citizens of them? And what is our task in this? This book shows that these questions are still very relevant and also particularly apply to young children. Children still enjoy playing outside, taking part in sports and playing with friends the most. We, nevertheless, see that they start using the internet at an increasingly young age and spend a growing amount of time online. Here too they play and talk to their friends. Dutch children are even front runner in Europe compared to their peers abroad. The number of children aged between 6 and 12 who own a mobile phone is also increasing rapidly.

As the focus of scientific research to date focused on teenagers, we knew relatively little on what 6- to 12-year-olds do with new media. But we are not completely in the dark. We have in this book identified all we know about 6- to 12-year-olds and supplemented it with new research. What has this resulted in? Below can be found an overview of the trends that have become apparent, the conclusions that can be made and the recommendations that follow on from there.

WHAT IS ON OFFER AND ITS USE

At increasingly younger ages, children make use of the internet and other new media. It often starts with easy games but it quickly expands into a versatile use. Within this diversity two clusters of applications stand out, namely, games and communication.

Children play on game consoles such as Playstation, Xbox or Wii, on the PC or laptop (with or without an internet connection), and on mobile devices such as *Nintendo DS* and *Playstation Portable*. Children who are online, almost all play casual games on the internet: small, usually free of charge, games, e.g. Bejeweled or Bubble Trouble. The selection of products on every platform mentioned is growing (see Chapter 2 – Games and Chapter 3 – Casual games).

The online product range for young children is growing steadily. Television channels such as Nickelodeon, Disney XD (previously Jetix.nl) and Z@ppelin cater to young children on the internet, fitting in with what they offer on television. Disney XD refers to the virtual world Dofus.nl and the game Online Soccer Manager, Nickelodeon to Spongebob.nl and Neopets. Nickelodeon, part of MTV Networks, even devises concepts for advertisers, e.g. Dr. Oetker (www.pauladekoe.nl and www.hieperdepiepgame.nl). More advertisers want to directly aim their ads at children via the internet, e.g. glue manufacturer Pritt using www.knutselwereld.nl, a site with arts and crafts tips for children.

This increase in product range is also expressed in games and the advance of virtual worlds for children. The German Panfu has now also been made suitable for use in the Netherlands, for children from 4 years onwards. The Swedish Stardoll has also been made accessible for Dutch girls. Disney's Club Penguin, for children from 5 years onward, is not yet available in Dutch, but that will not take long. In 2007 Disney acquired Club Penguin for USD300 million and that has to be recouped. Chapter 4 (Virtual worlds) charts the growth in these environments and deals, in particular, with the popular Habbo.

Parallel to the increasingly connected audience of children the product range especially produced for them is rising. Children are becoming big business. The virtual worlds are, as a rule, freely accessible, but if you really want to have fun you have to take a subscription (for approximately €5 per month) or you spend money on credits so that you can buy things that make your stay and contact with others more appealing. In virtual worlds, such as goSupermodel, children do not only see advertising banners and billboards, they are also directly urged to spend money on virtual products such as clothing, shoes and accessories to dress their personal avatar. This is called product placement. Casual game portals such Spele.nl and Speeleiland.nl earn a lot of money on the advertisements shown before the game is played. In advergames the game itself is the advertising expression. Children are confronted with various types of advertising from a young age.

Commerce goes even further. Children are used as 'brand ambassadors' and sometimes even make the advertisements themselves. Jumbo (manufacturer of board games) urged children to make their own Stratego commercial via the Jetix website and send it in. The winner would be shown on Jetix television. And on goSupermodel girls were called on to advertise for goSupermodel everywhere on the internet. Charities also use children as 'brand ambassadors', e.g. World Wildlife Fund that encouraged children to campaign for the shark. They could log in and make their own virtual island. If they sent adults a donation request they could earn extra points for their island.

In short, an increasing number of sites are directed at young children, often on commercial grounds, and the use by children is growing proportionally.

Many children feel attracted to offers not actually meant for them. The social network sites stand out the most. Hyves was originally set up for students but is now also embraced by teenagers and young children. One-third of Dutch children aged 8 already have a Hyves profile (see Chapter 6 – Hyves). For children from 8 years onward, if they learn bit by bit to communicate via new media, the social possibilities make the new media even more appealing. This is shown by the use of Hyves and MSN and by participation on sites with a strong social component. Hyves is now more popular than MSN (in 2009 20% of the children aged between 6 and 9 used MSN and 39% used Hyves. Of the 10- and 11-year-olds, 54% used MSN and 69% used Hyves[1]).

The opportunities of keeping in touch with friends are growing even further as more children's sites are incorporating social network elements. On Spangas.nl, the site of the Dutch television programme of the same name, children can promote themselves with a personal 'agenda'. Virtual worlds were initially places where they could walk around with their avatar but now sites such as Habbo and Stardoll have also become social networks. Children can make a profile page on which they can display their network of digital friends.

Then there is also the spread of mobile phones. This is still ongoing. Of the 6- to 10-year-olds approximately half now have a mobile phone and among the 12-year-olds, nearly all have a mobile phone. This is quite something. Online advertisements used to only spur children to go to the shops and most parents had a grip on this. But now children with a mobile phone have a new purse, with prepay credit as the new currency. Voting for your favourite Idols candidate, buying furniture on Habbo or clothes on Stardoll, text messaging via MSN of ordering a love test for your mobile phone – marketing professionals can approach children directly and persuade them to paying directly.

SKILLS

In order to be able to function online, children need to learn digital skills. This includes the skill of being able to find your way in matters usually meant for adults. Part of this product range, however, has a particularly great appeal on children although they are not up to this yet. In Chapter 9 (Information skills) the author shows that young children are nowhere near as handy as is often thought. They hardly read anything online and click endlessly until they have found what they were looking for, during which much can go wrong. They make typing mistakes and end up on typo sites such as 'nikkelodeon.nl' or 'sesamstaat.nl', websites referring to misleading texting services or sometimes even hard core porn.

This does not alter the fact that they really want to learn and are not afraid to make mistakes. Children learn some digital skills with the greatest of ease. Casual games and MSN Messenger don't call for extensive training. It is, however, a misconception to deduce that the quickly acquired skills in the area of gaming and communication also mean that children can also quickly and easily master other skills. Information skills in particular (search, find, assess, process, etc) are not automatically mastered. They have to be taught.

What can and cannot be said and what you can and cannot show in online communication is also something that has to be learnt. Learning these skills and social skills is accompanied by the risk of cyberbullying and sexual harassment when the children grow older. In learning these skills children could do with some help from each other as well as from parents and teachers.

The extent to which they miss these skills is shown by the difficulty children have in recognising advertising in new media (see Chapter 7 – Advertising, and Chapter 9 – Information skills). But it has not been made easy for children. The conflict of interests of commerce and entertainment make it difficult to distinguish advertising from non-commercial messages.

New media offer children all kinds of opportunities to entertain themselves and to be in contact with friends but this can also be coupled with risks. Many children are, for example, confronted with violent or pornographic images.[2] But children also encounter other content that could be harmful or at the very least undesirable, e.g. in games (frightening scenes, expressions of a sexual nature, malicious joy, etc.).

In addition, children are often the victim of misleading commercial practices. Although advertisers have recorded regulations for advertising in new media (in the Netherlands, among other, in the Reclame Code – Advertising Code) it is particularly in the digital children's domain that these regulations are violated. The text messaging services for which they have unwillingly obtained a subscription are expensive. The telecommunications sector is sharpening its own rules of conduct and the Ministry of Economic Affairs is creating additional legislation to prevent problems. This should mean that the texting service problems are finally a thing of the past. But this is not all. The Dutch Consumer Authority is now focusing its attention on the misleading commerce in social networks such as Hyves. Children are approached by model agencies who promise them mountains of gold but after an expensive photo shoot never contact them again.

Participation in social networks and other online communication has risks. It can bring about contact with strangers whose intentions are not always good – not just crooks but also pedophiles. By putting their own information on the web, children also give away personal information that they could better not have. Children should learn that all information (e.g. addresses, telephone numbers and photos) put online can be seen by the whole world and as such also stays there forever. Annoying photos put online can have far-reaching consequences of which young cyberbullies are usually not aware.

Copyrights can also lead to problems. On the one hand, children are unaware that content 'publicly' placed on the internet also belongs to others. And on the other hand, they are upset if their belongings are stolen. Children cut and paste others' material with all the consequences. In 2009, a 15-year-old boy had to pay thousands of euro in damages to a photographer because he had placed a photo of Johan Cruijff on his site. And in Habbo it is no exception that digital furniture (worth a lot of money) is stolen.

The fact that children still make use of the new media in large numbers despite these risks, shows that they (and their parents) are not deterred by obstacles. The advantages amply compensate for the disadvantages. The rapid spread of new media, the growth of the digital products on offer and the increasingly more social nature of many services have given children many more possibilities to entertain themselves and to learn. Some children read a book and chat to its author. Other children watch entire nature films on YouTube. New media are also encouraged by school. Pupils are given assignments to use the internet for school projects. And shops have software that children can use to practise school material in the weekends and during the holidays.

Some parents are less enthusiastic with the ever increasing social nature of the internet. But it is particularly this social aspect, expressed as contact via MSN, the use of social networks and playing multi-user games, that appeals to the children. And it is useful. In Chapter 5 (On-line communication) we saw that it contributes to their social, emotional and cognitive development. Children are taking the first step to shaping their own identity and acquiring a place in the social network of peers and others. The use of new media plays an important role in gaining self-confidence, social and emotional support and maintaining friendships.

12.2 • CONCLUSIONS
MAKE RESILIENT AND PROTECT

This book shows that we need to make children resilient and we need to protect them. It is not one or the other. In particular, the industry, education system and parents are responsible. They are equally responsible. How can they take their responsibility to allow children to get the most out of new media, in a responsible way?

Parents are mainly positive when it comes to the use of computers as long as it does not take too much time or is at the expense of school achievements. Most parents feel that the internet is good for the overall development of their child and that they acquire useful information and learn new things from the internet.

Many parents are committed media educators and apply different strategies to mediate in their children's use of new media. There are also parents who omit any kind of guidance. Consciously or unconsciously they trust that everything will work out well. And this is usually the case, but not always.

The arrival of new media (the internet and mobile phones) have changed the role of parents tremendously, according to Peter Nikken and Justine Pardoen in Chapter 11. Parents themselves grew up in an era in which access to media was limited, but even more important: in those days their parents could to a great extent decide on what their children could and could not see. This has now become much more difficult. Media education has become much more complex.

Parents from higher social classes are generally more active in media education than parents from lower social classes while most of the vulnerable children come from the latter social class. They are not only relatively often the victim but also the offender. It is particularly their parents who need support. They could well use the help of content providers whereas education especially has a role where parents have not fulfilled their task.

The gaming industry

In the Netherlands, much experience has been gained from the classification of films and television programmes by way of the Kijkwijzer. Icons are used to indicate below which age the content could be harmful and other icons indicate the type of undesirable content it may contain (e.g. sex, violence, racism). To a certain extent, a similar system exists for games: Pan-European Game Information (PEGI). It was introduced in 2003 to classify games according to age to help parents in purchasing suitable games for their children. Almost all games now have a PEGI label. Large game producers such as Sony, Microsoft and Nintendo support PEGI. Games with a hefty content (violence, but also sex, discrimination, bad language, frightening images, drugs or gambling) receive a 12, 16 or 18 rating. These hefty games are, however, limited in number.

Things are different for games on the internet. The number of providers is much larger and more diverse. As a supplement to PEGI, the European Commission developed PEGI Online (PO). Producers and licensees can indicate on their website whether a game has a PEGI rating and the games' possible dangers. Unfortunately, PEGI Online can hardly be found on casual game portals, where children come across games that could be harmful for children younger than 18 years of age according to the PEGI rating. Many games with a violent content can be found on casual game sites and they are also easily accessible. It is possible that children aged 6 could end up on a site with a collection of different games resulting in them lying awake all night. If they play games with much realistic violence and blood they are not clearly warned.

Internet
Social media do not keep up with the rapid growth of young children active online. Much popular content is developed by producers who know how to appeal to children but are not aware of how to set up a sound moderation policy or do not know what to do with the flood of emails from children and parents. They learn this along the way by doing, when parents complain or when 'bad publicity' compels the producer to adapt the site.

Social network sites could show more consideration for their youngest visitors by filtering advertisements not suitable for a young group and by informing them more about privacy.

Also content made by professional, long-standing producers leaves things to be desired. Many sites for children are not user-friendly. They claim to be for children but children who can hardly read and just keep clicking do not find what they are looking for. Children do have to learn how to search, at home and at school but sites should also take the surfing behaviour of young children into account. Also by not showing them advertisements not intended for them. Making children's sites is an art in itself and an art still in its infancy. The internet as a medium is young and the internet for children is just out of the cradle.

Children should already learn to orientate themselves on the internet at primary school so that they can find what they are looking for and can filter the information found according to relevance and reliability. Teachers should not just possess basic skills in dealing with new media but they also need to know how they can incorporate new technologies into their lessons.

There are, however, still large differences between schools in the degree to which they apply ICT in their lessons and support the information skills of pupils. In summary: primary schools do have computers and digital blackboards but a view on the content is still lacking – how to go about with the devices and how to link their use to the curriculum.

In the Netherlands, the ministries of Education, Culture and Science and of Youth and Family encourage media literacy and the Ministry of Economic Affairs is attempting to improve the e-skills through the programme Digivaardig & Digibewust (Digi skilled & Digi aware).

12.3 • RECOMMENDATIONS
MEDIA LITERACY SHOULD GET A MORE PROMINENT POSITION IN EDUCATION
Make children digitally literate, start at the primary school
- Teach them to use basic and other digital techniques and applications properly (e.g. email);
- Teach them to actively search and to select and assess information on reliability and relevance;
- Teach them to use the correct digital means to communicate with others, to work together and to make content.

More focus on digital citizenship
More attention should be paid, at home and at school, to what makes someone a good citizen when using the internet, mobile phone and other digital media. We want to prevent cyberbullying among children and we want them to adhere to the law, also with new media. But children should also be told what they can do: what is good behaviour? Do you break up In Real Life, or can you also do this via your mobile phone? What makes you a good friend on the internet? In short, ethical behaviour.

Stimulate digital creativity

New media offer great opportunities for creating content. Nowhere near all children can do this. If children are capable of making and publishing content they are also more capable of reflecting on the media messages of others.[3] Stimulate children's creativity. Encourage them to develop games, to make websites and keep a weblog, on the condition, of course, that they do it safely. Media literacy does not necessarily have to be a separate subject at school. Schools can, for example, ask children to hand in their school project as a film rather than on paper.

EMPOWER PARENTS AND TEACHERS

Inform parents, teachers and professional carers

With the development of social networks and Web2.0 applications, new opportunities and risks have been created which require an additional round of public information. It is important that parents and teachers be informed on the digital technologies and how to use them responsibly. It is particularly important that they know how to deal with information found on social network sites, weblogs and photo sites so that they can discuss this with their children. Information for professionals, e.g. pedagogics, psychologists and juvenile court judges, would also be useful.

Help parents with good 'tools'

Get the government and the industry to create tools to enable parents to fulfil their task as supervisor in a way they want to. Tools with which they are informed about media (as now developed in Kijkwijzer and PEGI), but also tools to restrict the access to the internet and mobile internet.

A children's browser, MyBee (mybee.nl) developed by KPN, can be downloaded free from the internet, but for mobile internet there are no possibilities yet to block undesirable content. Children have free access to the internet via their mobile phone and parents are not being given the opportunity yet by the industry to restrict access. Give parents the choice between complete and restricted access when they enter into an agreement for their children in a shop. Enter into text messaging service contracts or not? Companies should not see this as a responsibility that puts them under pressure but as a good opportunity to improve customer satisfaction.

Train teachers

It is essential that teachers understand what it is about new media that appeals to children and that they learn to guide these children in technological and digitally ethical areas (which applications there are and how to use them in a responsible way). In addition, as a result of the digital applications teachers have to consciously think about separating their work and their private life. Are you as a teacher allowed to give homework assistance after 10:30 pm on Hyves? Teach teachers how to deal with complaints about cyberbullying. There are many inspiring teachers and ICT coordinators who are experts in this area but a large group of teachers have much to learn. The advancement of knowledge in these areas should become less noncommittal. Librarians already play an important supporting role.

PROTECT CHILDREN MORE EFFECTIVELY AGAINST RISKS

Steer children away from harmful content

Further support by PEGI Online (PO) is important to help parents make choices. This approach has to be international as online games do not take country boundaries into account. The European Commission can fulfil a leading role but the participation by national organisations such as NICAM (Netherlands Institute for the Classification of Audiovisual Media) are of vital importance in order to involve as many game producers as possible in PO. This also applies to the producers of casual games. If games have a PEGI icon it becomes easier for children and parents to click away if a game has a high age rating.

Improve protection of children against online commerce

There is room for improvement in the cooperation between the government, producers and advertisers with the aim to ensure a sound commercial new media environment for children, with a more pro-active supervision of compliance with agreements and codes made.

As such, combating misleading advertising for children should be not be a problem, also on the internet. There are, after all, the Wet op Oneerlijke Handelspraktijken (Unfair Commercial Practices Act) and the Reclame Code (Advertising Code). But developments are so rapid that the industry and the government do not appear to be able to keep up the pace. Clients make advertising producers approach the boundaries of the indecent and bind children to them in a way that is certainly against the Reclame Code.

In 2009, Jan Driessen of the Dutch Association of Advertisers (BvA) called for 'socially responsible communication'. Increasingly, the industry takes a collective position and calls for advertising colleagues to make 'honest advertising'. This is a positive trend. Large advertisers are following this appeal but there are still advertisers who are not bothered, although this is sometimes because they are not aware of the rules. But let us not forget the site operators. Sites such as Hyves have a policy to guard young children from advertising not suitable for them. For years, misleading text messaging adverts could be found on all sorts of sites without site operators objecting. In fact, it was a good source of income for them. A lot could be said for updating the Reclame Code. There should be more focus on making advertising for children easier to recognise, in whatever medium. On the basis of scholarly views on what children at certain ages can and cannot recognise as advertising, a code could easily be drawn up, more in line with the advertising children come across nowadays. Also guidelines should be included that can assist children in recognising advertisements, e.g. the location of an ad or the way in which it is shown. But this is not enough. Sites too should take note of Jan Driessen's appeal and take the youngest users into account. They should make more effort to avert advertising that is against the law and in conflict with the children's and adolescent's advertising code. In addition, it is important to get a better understanding of the way in which children process advertising and which elements and characteristics attract them. The Reclame Code could be sharpened on the basis of these insights so that certain advertising strategies, against which children cannot defend themselves, would be restricted.

Take children to high-quality content
To help parents keep their children away from harmful texts and images we could put more effort into 'positive' content. It would seem to speak for itself that the more suitable content for younger children there is on offer, the smaller the chance that they get to see harmful content. Simply because when children are attracted to positive content they will not go looking for something else and by mistake come across harmful material.[4]

Content providers also benefit from guidelines for high-quality content for children. Example: how can you earn from a good website for children without overloading it with commerce? How do you set up a good moderation policy so that visitors can communicate with each other safely? Even just a discussion on these topics would help content providers to scrutinise what they themselves offer and improve where possible.

Relatively little is known on the use of new media by young children when compared to that of teenagers. A number of questions will remain unanswered even after reading this book. Some of the most burning issues are:

- **Relation between risks and danger:** to what extent do risks actually lead to danger?
- **Positive effects**: to what extent do different types of use have a positive effect?
- **Evaluation research**: to what extent are policy interventions in the area of children and the use of new media successful?
- **Media literacy:** developing a good measuring tool has high priority. Only then can it be determined whether media literacy has the desired effects.
- **Parental mediation**: carry out additional research on the relative success of the different types of parental mediation.
- **Context of media use:** study the use of new media in the context of other leisure activities and in a social context with friends, parents and teachers.

NOTES

1   Sikkema, P. (ed.) (2009). Jongeren09. We laten ons niet gek maken.
    (Youngsters09. We won't let them make a fool of us.) Amsterdam: Qrius.
2   Hasebrink. U., Livingstone, S. & Haddon, L. (2008). *Comparing Children's Online
    Activities and Risks across Europe: Cross-national comparisons for EU Kids Online.*
    London: EU Kids online (Deliverable D3.2). Available at www.eukidsonline.net.
3   Duimel, M. (2009). *(Onbewerkt) Beroemd* (Rough version) Famous. The Hague:
    Stichting Mijn Kind Online (My Child Online Foundation).
4   Livingstone, S. (2009). A rational for positive online content for children.
    Communication Research Trends, 28(3): pp.12-17.

About the authors

Lotte Boot

Lotte Boot (1976) is a freelance journalist. She is a regular contributor to the Stiching Mijn Kind Online (My Child Online Foundation) and writes articles for magazines such as MIND Magazine, Margriet and Amnesty's 'Wordt Vervolgd' ('Will Be Continued'). She started her career at the HP/de Tijd magazine, after which she became senior news editor at Planet.nl. One of her projects involved a series of interviews with child porn investigators of the KLPD (Dutch National Police force). Until 2007, Lotte lived in the US and first started writing for and about children in her Kidsplanet.nl column 'Lotte in America'.

Alfons ten Brummelhuis

Alfons ten Brummelhuis is head of research at Kennisnet (public knowledge centre) and, prior to that, worked for the University of Twente. He is in charge of a research programme looking into the use of ICT in education. Alfons is also one of the authors of the Vier in Balans Monitor (the 'Four in Balance Monitor'): an annual review of the actual state of affairs on ICT use in education. He is particularly interested in the knowledge development on what works in the educational use of ICT and what does not.

Menno Deen

Menno Deen (1982) is a PhD student at the Fontys Hogeschool for ICT (University of Applied Sciences) in Eindhoven. He carries out research on motivations for playing and learning, under the supervision of Prof Ben A.M. Schouten. Menno obtained his BA degree Design & Technology from the Hogeschool van de Kunsten Utrecht (Utrecht School of the Arts) after which he completed an MA New Media and Digital Culture at the University of Utrecht. Following his studies, Menno worked as researcher for Ranj Serious Games, before taking up his PhD in Eindhoven.

Marion Duimel

Since 2005, Marion Duimel (1976) has been working for the Netherlands Institute for Social Research, SCP, where, together with Jos de Haan, she wrote 'Nieuwe links in het gezin' ('New links in the family') about the digital world of teenagers and the role of their parents. In addition, she takes on freelance assignments in photography, video editing and research projects, among other for the Stichting Mijn Kind Online (My Child Online Foundation). By combining research and creativity, Marion tries to capture reality in a wide perspective.

Jos de Haan

Jos de Haan (1960) is head of the Tijd, Media en Cultuur (Time, Media and Culture) research Group of the Netherlands Institute for Social Research, SCP and, since 1 September 2006, he is Professor in ICT, Culture and Knowledge Society at the Erasmus University in Rotterdam. Together with Marion Duimel, Jos wrote the SCP report 'Nieuwe links in het gezin'('New links in the family') about the digital world of teenagers. With Christian van 't Hof he co-edited the Jaarboek ICT and Samenleving 2006 (ICT and Society Yearbook 2006), about the digital generation.

Jeroen Jansz

Jeroen Jansz (1958) is Professor of Media and Communication at the Department of Media and Communication (Erasmus University Rotterdam). He lectures in the International Bachelor Communication and Media (IBCom) and the MA programme Media, Culture and Society. His research is about the way in which media consumers use Internet applications (e.g. YouTube, Flickr or blogs) to become their own generator of media content. The project is a continuation of his research on video games and (young) players, which he worked on at the

University of Amsterdam during the past decade. For details about his publications see: http://www.fhk.eur.nl/personal/jeroen_jansz/bio/

Nathalie Korsman

Nathalie Korsman (1973) was editor-in-chief of the Dutch Habbo Hotel (www.habbo.nl) from its launch in 2004 until 2009. At present she is freelance web editor, specialised in the responsible design of commercial and non-profit websites aimed at children and adolescents. In addition, Nathalie is a member of the editorial team of the MyBee children's browser and the committee of recommendation of the stichting e-hulp. nl (e-help foundation).

Els Kuiper

Since May 2010, Els Kuiper (1958) has been researcher and lecturer at the Department of Pedagogical and Educational Sciences of the University of Amsterdam, and prior to this worked at the VU University Amsterdam. She graduated in educational science and for a number of years has been doing research on the use of the internet in education and the way in which education can contribute to the critical use of the internet by children. She has written about this subject in various scientific journals and general interest magazines and in 2007 published the PhD thesis *Teaching Web literacy in primary education*.

Peter Nikken

Peter Nikken (1960) is senior at the Nederlands Jeugdinstituut (Netherlands Youth Institute). He graduated as a psychologist on research for Sesame Street in 1986 and has since then been a leading expert in Youth and Media in the Netherlands. In 1999, he received his PhD at the Leiden University following a thesis on Quality standards for children's TV programmes. Peter is a member of several national committees, among which, the science committee of the NICAM (Netherlands Institute for the Classification of Audiovisual Media), the advisory board of Stichting Mijn Kind Online (My Child Online Foundation) and MediaSmarties, a project that should lead to a rating system for suitable children's media. Over the past few years he has published several scientific papers and general interest texts on subjects concerning youth, media and parental mediation. His most recent book Mediageweld en Kinderen ('Media violence and Children', 2007) summarizes the scientific output from the last decade on media violence.

Justine Pardoen

Justine Pardoen is editor-in-chief of Ouders Online (Parents Online, www.ouders.nl), the online journalistic platform for parents. Together with KPN (KPN Royal Dutch Telecoms), Justine founded the Stichting Mijn Kind Online (My Child Online Foundation), to help parents and professionals with media education. She gives presentations about 'sex in the media' and other aspects of the digital youth culture and wrote various books on internet education. With Remco Pijpers she maintains a stimulating weblog on these themes. Justine also keeps a weblog about children and games, for the website Weet Wat Ze Gamen (Know What They're Gaming), an initiative of the government funded programme Digivaardig & Digibewust (Digi skilled & Digi aware).

Jochen Peter

Jochen Peter (1972) is Associate Professor at the Amsterdam School of Communication Research (ASCoR) of the University of Amsterdam. As member of the research team Jeugd en Media (Youth and Media, see www.ccam-ascor.nl), Jochen focuses on the consequences of internet use on the socio-psychological and sexual development of adolescents. He has written more than 50 academic articles.

Remco Pijpers

Remco Pijpers (1969) is co-founder and Director of the Stichting Mijn Kind Online (My Child Online Foundation). He is responsible, among other, for the children's browser MyBee and the Gouden Apenstaart (award for the best children's website). He also coordinated the youth campaign Internetsoa.nl (on sexually transmitted diseases). With Justine Pardoen of Ouders Online he wrote a number of books on internet education, among which 'Verliefd op Internet'('In love with the Internet'), about the internet behaviour of adolescents. With Carry Slee (well-known Dutch writer of children's books), he wrote the children's book 'Kindertelefoontips' (Child helpline tips). Remco is a journalist and historian.

Esther Rozendaal

Esther Rozendaal (1981) works as a PhD student at the Amsterdam School of Communications Research, ASCoR, of the University of Amsterdam. Her research focuses on whether being advertising-literate can reduce the sensitivity of children toward advertising effects. In addition to her work as a PhD student, Esther Rozendaal also works as a scientific advisor, specialised in consumer behaviour of children.

Mijke Slot

Mijke Slot (1979) is researcher at TNO Information and Communication Technology in Delft and university lecturer at the International Bachelor Communication and Media at the Erasmus University Rotterdam. Mijke is writing her PhD thesis on the changing relationships between consumers and producers in the online media entertainment domain.

Patti M. Valkenburg

Patti M. Valkenburg has been Professor of Youth and Media at the Department of Communication Science of the University of Amsterdam since 1998. She is the Director of CCAM, the Center of Research on Children, Adolescents and the Media (www.ccam-ascor.nl). Professor Valkenburg is a member of many national and international boards and committees, including the science committee of the NICAM (Netherlands Institute for the Classification of Audiovisual Media), and the social science division of NWO (Netherlands Organization for Scientific Research). She wrote more than 130 articles and book chapters, and received more than 20 awards for her scientific work.

Contributors

The book Contact! is an initiative of the following three organisations:

DIGIVAARDIG & DIGIBEWUST

Digivaardig & Digibewust ('Digi skilled & Digi aware') is a Dutch programme set up by the government, trade and industry and social organisations. The aim of the programme is to reduce the number of people who lack or have insufficient digital skills and to stimulate the responsible, aware and safe use of digital means. The specific target groups are: the digitally illiterate, adolescents and their carers (parents and teachers), employees in small and medium sized companies and senior citizens. The programme is supported by the Ministry of Economic Affairs, the European Commission and various companies (KPN, UPC, Microsoft, NVB, IBM, NVPI, SIDN). The programme is being conducted by ECP-EPN, the platform for the information society.

- email: info@digivaardigdigibewust.nl
- website: www.mijndigitalewereld.nl

THE NETHERLANDS INSTITUTE FOR SOCIAL RESEARCH

The Netherlands Institute for Social Research (SCP) is an interdepartmental academic institute established by Royal Decree on 30 March 1973. On 1 January 1974, the agency started. The SCP carries out independent research and reports to – both solicited and unsolicited – to the government, the Senate and the House of Representatives, ministries and other social and governmental organisations. In addition, the research results are also used by professional and administrative executives in the quaternary sector, by science and by individual citizens. SCP reports are available as hard copies and can be downloaded free of charge from the website. The SCP employees contribute to public and scientific debate through their articles, lectures, contributions to conferences and interviews.

- email: info@scp..nl
- website: www.scp.nl

MY CHILD ONLINE FOUNDATION

My Child Online Foundation is an independent expertise centre for youth and media. The foundation's main aim is to help parents, schools and professional carers with internet education. This is achieved, among other, by organising parent evenings and courses. My Child Online's spearhead is quality of digital media for children – My Child Online stimulates a good digital product offering by, among other, the annual Gouden @penstaart award for the best children's website and MyBee, a free children's browser (www.mybee. nl). In addition, My Child Online is responsible for the project management of 'children and new media' for, among other, the police and gives advice to companies and organisations on communicating with children via the internet.

- email: redactie@mijnkindonline.nl
- website: www.mijnkindonline.nl